THE
BERLIN IRON BRIDGE CO.
EAST BERLIN CONN.

LANDMARK AMERICAN BRIDGES

ERIC DELONY

CHIEF, HISTORIC AMERICAN ENGINEERING RECORD

NATIONAL PARK SERVICE

U.S. DEPARTMENT OF THE INTERIOR

AMERICAN SOCIETY OF CIVIL ENGINEERS
345 EAST 47TH STREET, NEW YORK, NEW YORK 10017-2398

Bridges remain a timely and provocative subject not only in the history of technology and historic preservation, but in engineering, transportation, and urban planning as well. As the single most important artifact of rail and highway networks, a bridge's historic significance is critical to understanding the role it plays in America's engineering and transportation history. This book features over 200 images of approximately 90 American landmark bridges arranged chronologically so that the evolution of American bridge building is revealed. These images are from the collection of the Historic American Engineering Record (HAER) which was established in 1969 to create a graphic and textual archive of America's industrial and engineering achievements of historic interest. Through these drawings and photographs, bridges are shown to be testaments to engineering creativity, technological skills, and teamwork. Additionally, an introductory text, a timeline of bridge history, a bibliography, and a listing of bridges in the HAER collection enhance this visual history of bridge building in America.

Library of Congress Cataloging-in-Publication Data

Landmark American Bridges / by Eric DeLony
p. cm.
Includes index.
ISBN 0-87262-857-4 (ASCE)
ISBN 0-8212-2036-5 (Bulfinch Press)
1. Bridges—United States. 2. Historic bridges—United States.
3. Bridges—United States—Pictorial works.I. DeLony. Eric.
II. American Society of Civil Engineers.
TG23.L36 1992
624´.2´0973—dc2092-24530
CIP

Library of Congress Catalog Card No: 92-24530
ISBN 0-87262-857-4 (ASCE)
ISBN 0-8212-2036-5 (Bulfinch Press)
First Edition
Manufactured in Mexico

cover jacket: Coos Bay (McCullough Memorial) Bridge, Jet Lowe, 1990
HAER

*This book is dedicated
to the hundreds of "pontists" who share the passion
of saving the landmark bridges of America*

Many people have contributed to the production of this book, but I want to recognize several who were particularly helpful. First is the chief of the HABS/HAER Division, Robert J. Kapsch, who created the opportunity to write the book of my dreams. He played a pivotal role in helping me with the book's planning and organization.

Virginia Fairweather, Editor in Chief at ASCE, was also instrumental in the start–up and planning of the book. Zoe Foundotos, ASCE's Acquisitions Editor, saw the project through from beginning to end.

HABS Historian Caroline Bedinger did yeoman's service pulling the drawings and photographs from the Library of Congress collections. Jet Lowe, staff photographer, spent days and evenings in the darkroom processing more than two hundred 11 x 14–inch prints, each hand–printed with immaculate care and quality.

Critical reviews came from Tom Peters, Neal FitzSimons, and Robert Vogel. Tom Peters, a great scholar and authority on bridges currently at Lehigh University, diligently checked and rechecked the timeline for accuracy. Neal FitzSimons, founder and first chair of ASCE's Committee on History and Heritage, reviewed captions. Robert Vogel, former curator of mechanical and civil engineering at the Smithsonian Institution and my favorite critic, read and commented on both the timeline and picture captions. The accuracy checks they provided were a comfort to me as an author. Additional thanks and accolades go to the designer, Stephanie Schaffer, and ASCE's Production Manager, Shiela Menaker, Neil Gaffney, Peg Peterson, and Geoffrey Howard at ASCE, and especially to Sherry Kay Campbell.

Writing a book is an individualistic experience. The author is ultimately responsible for all the contents. It also is a collaborative effort whose success depends on the interest, enthusiasm, and assistance of many people. I've been fortunate to have the very best photographer do all the printing, the very best reviewers to ensure accuracy, and the very best editors to ensure readability. To these people, I am eternally grateful.

Eric DeLony
Chief, Historic American Engineering Record

INTRODUCTION

*F*ew would dream of tearing down the Brooklyn Bridge, the Golden Gate, or a wooden covered span, but the truly outstanding examples of our concrete arches, composite-cast and wrought iron trusses, steel trusses and movable spans are lost every day. We should, instead, equate the truly outstanding bridges with other U.S. landmarks. We should protect and rehabilitate them as we would Independence Hall, Monticello or Mount Vernon.

The U.S. Department of Transportation estimates that of 576,000 U.S. bridges, approximately 39 percent are structurally deficient or functionally obsolescent. Unfortunately, the same federal and state programs that help rehabilitate deteriorating highways often cause the destruction of historic bridges. With this book, I hope to identify these landmark bridges, to focus attention on those that illustrate the history of bridge building, transportation and engineering, and to encourage their preservation.

What is a landmark bridge? The Secretary of the Interior states that landmark bridges are those of exceptional value to the nation as a whole. They must illustrate or interpret the heritage of the United States in engineering, technology, transportation, industry, history or culture. Landmark bridges must also possess a high degree of integrity of design, materials, workmanship, setting, feeling and association. Nine percent of the bridges in the United States are eligible for listing on the National Register.

The Historic American Engineering Record (HAER) and the Historic American Buildings Survey (HABS) were established in 1969 and 1933, respectively, to create a graphic and textual archive of America's building arts—bridges, buildings, factories, churches and other historic structures. Together, these nationally valuable collections contain documentation on more than 1,000 American bridges. More than 900 bridges are documented in the HAER collection. The HABS collection documents the remaining 100. These archives, available to the public and housed in the United States Library of Congress in Washington, D.C., form the basis for this book.

Both the HAER and HABS programs function under an agreement among the Library of Congress, the National Park Service of the U.S. Department of the Interior, the American Institute of Architects and the American Society of Civil Engineers. A 1985 protocol expanded cooperation and support for the HAER program to include the American Society of Mechanical Engineers, the Institute of Electrical and Electronics Engineers, the Society of Mining, Metallurgical & Petroleum Engineers and the American Institute of Chemical Engineers.

It is important that we, as a nation, are aware of the types of common bridges that have been disregarded or overlooked even by preservationists and engineers. For instance, in a single generation we have nearly succeeded in eliminating one of the most important developments of the built environment—the prefabricated metal truss bridge.

The phenomenon of the metal truss symbolizes the fundamentally American values of entrepreneurialism, craft and unbridled invention and creativity. The metal truss helped Americans cross thousands of streams and rivers, reach new markets and create new businesses as the frontier moved west. Marking the desire to improve technology, hundreds of patents were granted in the 19th and 20th centuries. While many patents were granted to trained engineers, many went to the crafters—millwrights and mechanics. These were untrained "apple tree engineers," who recognized a need and sought engineering solutions that proved to be practical and sometimes a little bizarre. If not addressed now, these artifacts of the American landscape, both rural and urban, are threatened with extinction.

Fortunately, several programs have been set up to protect this engineering heritage. The current federal bridge program began with the collapse of the Point Pleasant Bridge over the Ohio River in 1967, in which 46 lives were lost. Following the work of a presidential task force, Congress passed legislation to enhance bridge safety. Landmark highway acts have since been passed, to establish bridge inspection, replacement and preservation programs. These include the Federal Aid Highway Act of 1968, the Federal Aid Highway Act of 1970, the Surface Transportation Assistance Act of 1978, the Surface Transportation & Uniform Relocation Assistance Act of 1987 and the Intermodal Surface Transportation Efficiency Act of 1991.

Historic bridge inventories have been completed in nearly all states, thanks to these legislative initiatives. But, sadly, relatively few historic bridges have been rehabilitated, relocated or reused since the federal bridge rehabilitation program began in 1978.

Several reasons account for the widespread destruction of historic bridges. These relate to technical problems due to deterioration and to the implementation of the programs established by highway legislation. These aspects converge in the current system used for rating the condition of historic bridges.

All bridges inspected under the Highway Bridge Replacement & Rehabilitation Program (HBRRP), established by the Surface Transportation Assistance Act of 1978, are assigned a point rating to assess their overall condition. Points are given for structural adequacy, the assessment of the structural capacity of the bridge; serviceability and functional obsolescence, which measures the geometric and traffic-capacity features of the bridge; and the necessity for public use, which is the assessment of the frequency of use and the importance of the bridge to the highway system.

Fortunately, bridges with unsatisfactory ratings are eligible for federal funds. These structures were typically built for less traffic and lighter loads than are found today, have been damaged by traffic or deicing chemicals or have received little maintenance. Even when they are not structurally deficient, many older bridges receive low marks for serviceability and are then rated as functionally obsolescent. Some of these bridges are capable of supporting modern traffic, but may have poorly aligned approaches by today's standards or overhead structural ties that offer insufficient truck clearance.

For any number of reasons then, historic bridges are sometimes eligible for rehabilitation funds or replacement.

With the aid of funds, some could be repaired or rehabilitated without damaging their historic character. This, however, has not always been the case. Nearly all projects using HBRRP funds require that assisted bridges meet the standards of the American Association of State Highway & Transportation Officials (AASHTO).

The 1991 Intermodal Surface Transportation Efficiency Act, known by its acronym ISTEA, is based on the flexible interpretation of the AASHTO standards and proposes to address positively environmental and preservation concerns in transportation planning. It is premature at this time to say how effective the new law will be.

Historic bridges can be preserved without endangering or inconveniencing the general public. Many historic bridges that do not meet AASHTO standards are perfectly serviceable. For example, AASHTO requires a 30-foot-wide deck. This width is impossible to achieve on many historic bridges built for narrower and one-lane roads. Yet one community, Allegan, Michigan, found an elegantly simple solution. After upgrading the structure, it made the bridge one way. Since other routes served the town, this solution worked well. Later, a traffic light was installed and two-way traffic resumed.

Historic bridge preservation is gaining momentum. The centennial of the Brooklyn Bridge in 1983 drew public attention to historic bridges, as did the 50th anniversary of the Golden Gate in 1987. There are environmental and quality-of-life reasons for saving old bridges as well. People are beginning to question the wisdom of having wider lanes and more efficient crossings that result in more cars and more development, rather than in lessening congestion and aggravation. Moreover, the staggering bill for repairs to the nation's road infrastructure in the years ahead demands alternatives to massive new construction.

Several state highway departments have initiated historic bridge programs. Where it has been tried, the repair of historic bridges has proven less costly than replacement. Ensuring the preservation of bridges, however, requires commitment. The local government must document the condition of the structure and the costs of rehabilitation or replacement. Such a study should also document the bridge's safety history, which is especially important for structures rated as functionally obsolescent while being structurally sound. Often the record shows few or no accidents, rendering arguments for replacement due to poorly aligned approaches or other "design inadequacies" difficult to sustain.

Preservation alternatives for historic bridges include continued use for traffic or conversion to a new use. Continued use for traffic may require that a bridge be dismantled so that the parts can be inspected. Sound pieces can be reassembled, while damaged or deteriorated pieces can be replaced. If the structural system is weakened, it can be reinforced at times, in such a way that the new members do not show.

In other cases of rehabilitation, bridges may require geometric modification, involving significant changes to the bridge. The disadvantage to geometric modification is that the rehabilitation can destroy the structure's overall historic character.

Other treatments might involve realigning the approach roads or changing the use of the bridge. Still other solutions change the way a bridge is used, by converting a span to one-way traffic, lowering load limits or physically moving it.

Keeping a bridge in vehicular use is not always possible. Alternatives include moving bridges to roads or trails that do not require full-service bridges, such as bike paths, hiking trails, or state and national parks. In some cases, a new bridge has been built beside the historic bridge, which is then left to pedestrians and fishermen.

Another important facet of this book is the reference information on historic bridges. The appendix presents a combined listing of the bridges documented in the Historic American Engineering Record and Historic American Buildings Survey collections. A time line appears throughout the book to highlight significant events, publications, patents, technological innovations and personalities in the evolution of bridge building. This time line will help the reader place the American experience in context with that of the rest of the world. It will document American bridge engineers' and fabricators' claim to a portion of world bridge building advances.

The book is arranged chronologically so that, as you turn the pages, the evolution of American bridge building will be revealed. This visual journey of photographs and drawings is accompanied by captions that place the bridges in historical context. It is my hope that this book will create a greater awareness of the historic bridges of the United States.

Few people willfully destroy historic bridges. Most are lost through ignorance. To preserve the historic bridges of the United States, we need the special cooperation of people in transportation with engineering expertise. Preservationists, architects and historians working alone will only partially succeed. Bridges are engineered structures. Their successful rehabilitation requires the ingenuity of engineers.

Along with the experts, we need an informed public who recognize the significance of historic bridges and care that the best examples are saved for the enrichment of posterity. If the majority of the bridges in this book are still standing in 20 years, then the book will have been successful.

EDL

FOREWORD

$\mathcal{T}$he American Society of Civil Engineers (ASCE) has long known the critical difference between individual effort and group effort. What one can do well, a group of two or more working together can do even better. In civil engineering, individual efforts create achievements. From group efforts come monumental achievements.

The teamwork approach has resulted in countless U.S. civil engineering structures that improved American living conditions during the past three centuries. It is also very much alive and evident in Eric DeLony's celebratory *Landmark American Bridges.*

Mr. DeLony's work pictorially documents American engineering's finest hour: its bridges. From another perspective, this publication is a most worthy addition to the collective fruits of a partnership spanning over two decades: the close interaction between ASCE and the Historic American Buildings Survey/Historic American Engineering Record (HABS/HAER).

As an engineer by vocation and a historian by avocation, I've always had a deep admiration for America's bridges. With the publication of *Landmark American Bridges*, my appreciation for these structures and the teamwork that created them has grown immeasurably. In *Landmark American Bridges*, these engineering marvels taken for granted for far too long are finally given their due. In page after page, through words, drawings and photographs, DeLony shows us the real significance of American bridges. It is not just that these structures served as the geographical workhorses of U.S. continental trade, but also that they've always displayed a decidedly different union of utility and artistry. The bridges in this collection are at once examples of engineering creativity, testaments to our technological prowess as a nation and legacies of our ability as a people to cooperate with each other.

As the Industrial Age of America progressed, hundreds of bridge patents were submitted. But before these patents were tendered, the inventors researched and sought out the advice and concerns of those groups for which they wished to contract services. Even as bridge builders embodied the entrepreneurial spirit of their time, they were pioneering a business approach that resulted in progress for people via a cooperative team attitude that took into account the needs of the railroads and communities they sought to connect.

The very idea of connecting in itself is what a bridge does. Likewise, connection through a common goal is what makes teams work. For results, one need only look to the great structures dominating our skylines that serve as reminders to that team spirit. As bridge engineer and former ASCE President James E. "Tom" Sawyer once told a group of civil engineering students, "The Brooklyn Bridge was not built by John Roebling. It was built by *a team* led by John Roebling."

Mr. DeLony's historical treatise on bridges is the latest in a long line of activities born in 1969 out of a coalition of concern on the part of three groups: ASCE, the Library of Congress, and the National Park Service. With the active assistance of these groups, HABS/HAER serves as the nation's chief historical gatherer, cataloger, and reminder of our industrial and architectural technological heritages.

For the civil engineering community, the importance of HABS/HAER is simple enough. The group's work documents and preserves civil engineering's greatest contributions to civilization, thereby providing a historical safe haven for the profession's betterment of our world.

As civil engineers practicing so close to a new millennium, we need to know from whence we came in order to plot where we will go. Thus, ASCE's keen interest in the HABS/HAER goals of documentation and preservation have made it a longtime partner on numerous projects of great historical significance.

As civil engineers, we benefit from having HABS/HAER archive the achievements of our profession through various publications and other activities. *Landmark American Bridges* is the latest HABS/HAER publication in an exhaustive series of works that justly frame the American industrial experience. These publications and activities serve, inform, and instruct, providing a continuity of purpose for a profession that gains confidence via its past achievements.

Mr. DeLony's collection should be taken as a warning signal. This collection tells us that the wholesale removal of that which is old and obsolete only jeopardizes our perception of our nation's vision of its industrial and social heritage.

This book is a bold step toward preventing that occurrence. If history is a discipline that helps those of us in the present understand the past a bit more clearly, then we owe Eric DeLony our heartfelt gratitude. Just as his subjects gracefully span physical space, Mr. DeLony's collection works to span time, helping us to reconcile the achievements of our past with the challenges of our future.

Edward O. Pfrang
Executive Director, ASCE

PREFACE

The idea for this book began at the opening of an American Society of Civil Engineers (ASCE) exhibit, organized by Curtis Deane, the Washington, D.C., manager of ASCE. It was Curtis, approximately twenty years after ASCE had helped found the Historic American Engineering Record (HAER) along the model of the older Historic American Buildings Survey (HABS), who began using HAER-produced drawings and photographs in ASCE exhibits. It was Curtis' idea to display the beauty of American engineering works through ASCE-produced exhibits. He used HAER photographs and drawings because of the very high quality of these images. And, we at HABS/HAER were very pleased to cooperate with Curtis and Susan Sarver in the production of those exhibits.

At this particular exhibit opening, I was talking to ASCE Executive Director Dr. Edward Pfrang. Ed and I had worked together years before at the Center for Building Technology, National Bureau of Standards (now the National Institute of Standards and Technology). We talked about how it was too bad that only a small fraction of the engineering profession, or indeed, the American public, would see these images of the grace and power of these engineering structures. We also talked about how each structure, particularly the innovative structures that somehow advanced our knowledge of engineering design and practice, represented a complete and interesting history in its own right. It was at that exhibit opening that we came to the conclusion that we needed to publish a book that could adequately convey the stories behind the design and construction of that beauty.

Eric DeLony, Chief of HAER, was a logical choice to write this book, select its illustrations, and otherwise oversee its development. Eric, an architect by training, developed his interest in engineering structures as a Fulbright Fellow in the United Kingdom almost twenty-five years ago. Upon his return to the United States, he found the ASCE, National Park Service, and Library of Congress struggling to develop a new organization to document outstanding examples of engineering—the HAER program. He became the first professional employee of that program and later its chief. He soon developed an overarching interest in the history of American bridge design and construction—not the romantic covered bridge of post colonial times but the metal truss and suspension bridges developed by an industrializing nation and designed by the new profession of civil engineering. For more than twenty years, Eric has been pursuing his interest in this aspect of our engineering heritage. During this time he has documented more than nine hundred outstanding examples of this engineering art many of which will be seen in these pages.

To implement the agreement between Ed Pfrang and myself, ASCE and HABS/HAER entered into a cooperative agreement. To oversee the publication process, ASCE appointed Zoe Foundotos, a very enthusiastic and energetic and thoroughly professional young woman.

This book is thus a cooperative effort between the American Society of Civil Engineers and the Historic American Engineering Record. It was ASCE that assisted in founding the HAER program, in 1969, to document America's engineering and industrial heritage. In subsequent years, HAER has created an archive of more than 2000 measured drawings, 34,600 large format photographs, and 2700 pages of history all housed at the Library of Congress, all processed for a 500-year archival life and all available for use by the public.

The HAER program is a unique blend of a public program supported by a professional society. The quality that HAER has achieved over the years is exceptional. This book provides only one small glimpse into the collection of documents amassed on America's engineering and industrial heritage.

Robert J. Kapsch
Chief, HABS/HAER

CONTENTS

Introduction... IV

Foreword..VII

Preface..VIII

Part 1: Turnpike, Canal & Railroad Bridges
            of the Pre–Civil War Era..................2

Part 2:   Civil War Era Metal Truss Bridges...............42

Part 3:   Era of the American Standard Bridge.........68

Part 4:   Great River Crossings.............................90

Part 5:   Modern Developments............................124

Appendix...146

Bibliography...151

Index..152

TYPICAL JOINT ASSEMBLY

Cast Iron Counter Diagonal

1 3/8" ø Wrought Iron

Cast Iron Joint Block

Obverse of Joint Block (reduced)

Wrought Iron Clamp

4" x 1 1/8" Wrought Iron Chord Bars

Wrought Iron Nuts

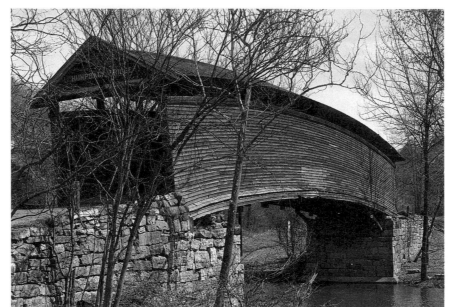

TIME LINE 1570–1828

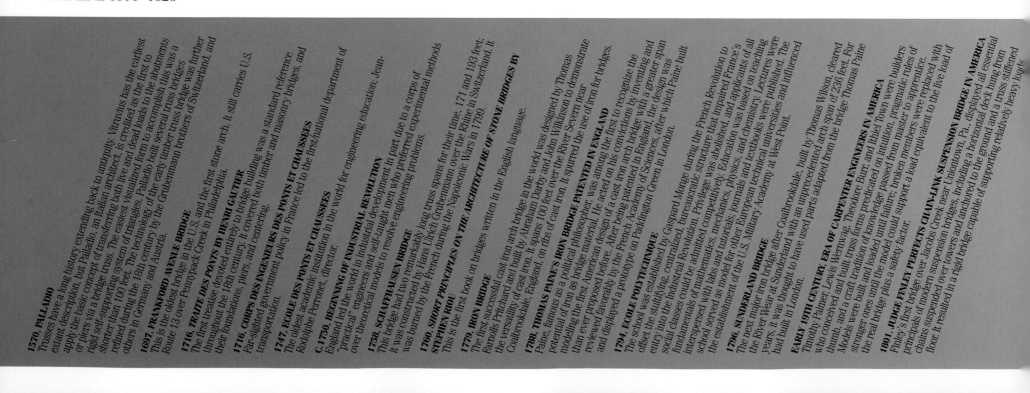

1570, PALLADIO
Trusses have a long history extending back to antiquity. Vitruvius has the earliest extant description, but Palladio, an Italian architect, is credited as the first to apply the basic concept of transferring both live and dead loads to the abutments or piers via a bridge truss. The easiest visualized form to accomplish this was a rigid self-supporting system of triangles. Palladio built several truss bridges, shorter than 100 feet. The technology of the early timber truss bridge was further refined during the 18th century by the Grubenmann brothers of Switzerland, and others in Germany and Austria.

1697, FRANKFORD AVENUE BRIDGE
This is the oldest bridge in the U.S., and the first stone arch. It still carries U.S. Route 13 over Pennypack Creek in Philadelphia.

1716, TRAITÉ DES PONTS BY HENRI GAUTIER
The first treatise devoted entirely to bridge building was a standard reference throughout the 18th century. It covered both timber and masonry bridges, and their foundations, piers, and centering.

1716, CORPS DES INGENIEURS DES PONTS ET CHAUSSEES
Far-sighted government policy in France led to the first national department of transportation.

1747, ECOLE DES PONTS ET CHAUSSEES
The oldest academic institution in the world for engineering education: Jean-Rodolphe Perronet, director.

C. 1750, BEGINNING OF INDUSTRIAL REVOLUTION
England led the world in industrial development through "practical" engineers and self-taught men who preferred experimental methods over theoretical models to resolve engineering problems.

1758, SCHAFFHAUSEN BRIDGE
This bridge had two remarkably long truss spans for their time, 171 and 193 feet; it was constructed by Hans Ulrich Grubenmann over the Rhine in Switzerland. It was burned by the French during the Napoleonic Wars in 1799.

1760, SHORT PRINCIPLES ON THE ARCHITECTURE OF STONE BRIDGES BY STEPHEN RIOU
This is the first book on bridges written in the English language.

1779, IRON BRIDGE
The first successful cast iron arch bridge in the world was designed by Thomas Farnolls Pritchard and built by Abraham Darby and John Wilkinson near Coalbrookdale, England. It spans 100 feet over the River Severn. It spurred the use of iron for bridges.

1788, THOMAS PAINE'S IRON BRIDGE PATENTED IN ENGLAND
Paine, famous as a political philosopher, was among the first to recognize the potential of iron as bridge material. He acted on his convictions by inventing and modeling the first American design of a cast iron arch bridge with a greater span than ever proposed before. After being patented in England, the design was reviewed favorably by the French Academy of Sciences, after which Paine built and displayed a prototype on Paddington Green in London.

1794, ECOLE POLYTECHNIQUE
The school was established by Gaspard Monge during the French Revolution to offset the stultifying, centralized, bureaucratic structure that impaired France's entry into the Industrial Revolution. Privilege was abolished, and applicants of all social classes could be admitted competitively. Education was based on teaching fundamentals of mathematics, physics, mechanics, and chemistry. Lectures were interspersed with labs and tutorials; journals and textbooks were published. The school served as model for other European technical universities and influenced the establishment of the U.S. Military Academy at West Point.

1796, SUNDERLAND BRIDGE
The next major iron bridge after Coalbrookdale, built by Thomas Wilson, cleared the River Wear at Sunderland with an unprecedented arch span of 236 feet. For years, it was thought to have used parts adapted from the bridge Thomas Paine had built in London.

EARLY 19TH CENTURY, ERA OF CARPENTER ENGINEERS IN AMERICA
Timothy Palmer, Lewis Wernwag, Theodore Burr, and Ithiel Town were builders who conceived and built truss forms predicated on intuition, pragmatic rules of thumb, and a craft tradition of knowledge passed from master to apprentice. Models were built and loaded until failure; broken members were replaced with stronger ones until the model could support a load equivalent to the live load of the real bridge plus a safety factor.

1801, JUDGE FINLEY PERFECTS CHAIN-LINK SUSPENSION BRIDGE IN AMERICA
Finley's first bridge over Jacobs Creek near Uniontown, Pa., displayed all essential principals of modern suspension bridges, including a horizontal deck hung from chains suspended over towers and anchored to the ground and a truss stiffened floor. It resulted in a rigid bridge capable of supporting relatively heavy loads.

Turnpike, Canal and Railroad Bridges of the Pre-Civil War Era

The earliest surviving bridges are stone, usually granite ashlar, random-coursed laid in lime mortar. Builders selected stone because of the material's durability. Most are of modest scale.

Not all masonry bridges were small. Some were built for a new revolutionary means of transportation for the early 19th century-railroads. These massive stone-arch structures, built in a classic Roman style, are called viaducts because of the great height of the piers and length required to carry the railroad over the river valleys of the eastern United States. Locomotives pulling heavy trains could not negotiate steep grades, so rather than descending into valleys only to climb out again, these stone viaducts maintained a nearly horizontal line. The Thomas, Canton, and Starrucca viaducts are the oldest surviving and three of the great monumental railroad structures in the country. All remain in service.

Covered wooden bridges also belong to the early decades of road building. The earliest date from the 1830s and '40s as none survive from the 18th century. Covered bridges have great appeal because of their nostalgic value and enjoy a large and aggressive following of enthusiasts watching out for their well-being. Many are remarkable works of engineering. For example, early concepts of pretensioning were introduced by the craftsmen who built them. European engineers visiting America during the early decades of the 19th century were impressed with our wooden-truss covered bridges. Most bridge patents granted during this period– Burr arch-truss, Town lattice, Long's truss, Howe and Pratt trusses–were for wooden-truss designs.

Two of the earliest metal-truss bridges are included in this section of covered bridges because they fall into this group chronologically and they show the evolution of truss design from timber to metal.

America's oldest surviving suspension bridges date from the mid-19th century, though the technology was available 50 years earlier. James Finley, a Fayette County, Pa., lawyer, was an amateur inventor and part of a group of men in the United States interested in the application of the principles of natural philosophy, science, and invention to practical matters, especially those concerned with internal improvements like bridges. He invented a chain-link suspension bridge in 1801 that displayed the major principles of modern suspension bridge technology.

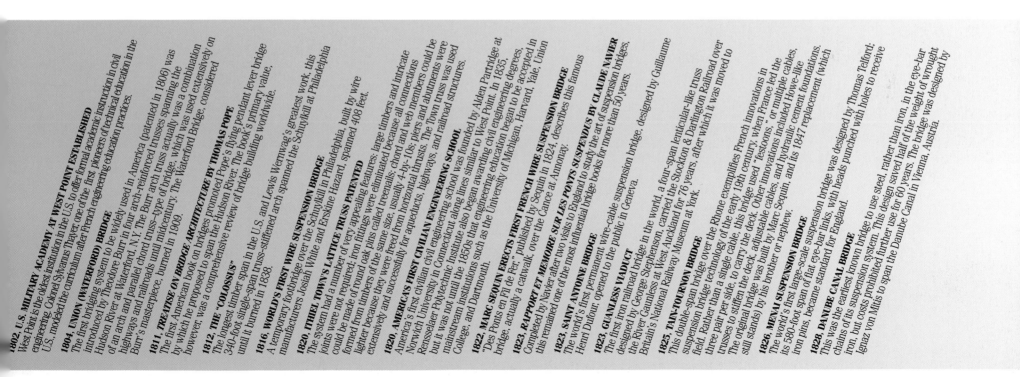

1802, U.S. MILITARY ACADEMY AT WEST POINT ESTABLISHED
West Point is the oldest institution in the U.S. to offer formal academic instruction in civil engineering. Colonel Sylvanus Thayer modeled the curriculum after French engineering education practices.

1804, UNION (WATERFORD) BRIDGE
The first bridging system to be widely used in America (patented in 1806) was introduced by Theodore Burr in four arch-reinforced trusses spanning the Hudson River at Waterford, N.Y. The Burr arch truss actually was a combination of an arch and parallel chord truss—type of bridge, which was used extensively on highways and railroads until midcentury. The Waterford Bridge, considered Burr's masterpiece, burned in 1909.

1811, TREATISE ON BRIDGE ARCHITECTURE BY THOMAS POPE
The first American book on bridges promoted Pope's flying pendant lever bridge by which he proposed to span the Hudson River. The book's primary value, however, was a comprehensive review of bridge building worldwide.

1812, THE "COLOSSUS"
The longest single-span in the U.S., and Lewis Wernwag's greatest work, this 340-foot single timber span truss-stiffened arch spanned the Schuylkill at Philadelphia until it burned in 1838.

1816, WORLD'S FIRST WIRE SUSPENSION BRIDGE
A temporary footbridge over the Schuylkill in Philadelphia, built by wire manufacturers Josiah White and Erskine Hazard, spanned 408 feet.

1820, ITHIEL TOWN'S LATTICE TRUSS PATENTED
The system had a number of very appealing features: large timbers and intricate joints were not required; iron fittings were eliminated because all connections could be made of round oak pins called treenails; chord and web members could be formed from timbers of the same size, usually 4-by-10s; piers and abutments were lighter because they were freed from horizontal thrusts. The Town truss was used extensively and successfully for aqueducts, highways, and railroad structures.

1820, AMERICA'S FIRST CIVILIAN ENGINEERING SCHOOL
America's first civilian civil engineering school was founded by Alden Partridge at Norwich University in Connecticut along lines similar to West Point. In 1835, Rensselaer Polytechnic Institute also began awarding civil engineering degrees, but it was not until the 1850s that engineering education began to be accepted in mainstream institutions such as the University of Michigan, Harvard, Yale, Union College, and Dartmouth.

1822, MARC SEGUIN ERECTS FIRST FRENCH WIRE SUSPENSION BRIDGE
"Des Ponts en Fil de Fer," published by Seguin in 1824, describes this famous bridge, actually a catwalk, over the Cance at Annonay.

1823, RAPPORT ET MEMOIRE SUR LES PONTS SUSPENDUS BY CLAUDE NAVIER
Completed by Navier after two visits to England to study the art of suspension bridges, this remained one of the most influential bridge books for more than 50 years.

1823, SAINT ANTOINE BRIDGE
The world's first permanent wire-cable suspension bridge, designed by Guillaume Henri Dufour, opened to the public in Geneva.

1823, GUANLESS VIADUCT
The first iron railroad bridge in the world, a four-span lenticular-like truss designed by George Stephenson, carried the Stockton & Darlington Railroad over the River Gaunless at West Auckland for 76 years, after which it was moved to Britain's National Railway Museum at York.

1825, TAIN-TOURNON BRIDGE
This double-span bridge over the Rhone exemplifies French innovations in suspension bridge technology of the early 19th century: when France led the field. Rather than bridge technology of the early 19th century, three pair per side, to carry the deck, this bridge used "festoons," or multiple cables, and trusses to stiffen the deck. Other innovations included Howe-like trusses, adjustable cables, and hydraulic cement foundations. The original bridge was built by Marc Seguin, and its 1847 replacement (which still stands) by his brother or nephew.

1826, MENAI SUSPENSION BRIDGE
The world's first large-scale suspension bridge was designed by Thomas Telford; its 580-foot span of flat eye-bar links, with heads punched with holes to receive iron pins, became standard for England.

1828, DANUBE CANAL BRIDGE
This was the earliest known bridge to use steel, rather than iron, in the eye-bar chains of its suspension system. This design saved half of the weight of wrought iron, but costs prohibited further use for 60 years. The bridge was designed by Ignaz von Mitis to span the Danube Canal in Vienna, Austria.

After the Frankford Avenue Bridge in Philadelphia (1697), Choate Bridge is the oldest existing stone bridge in the country. On the main north-south route of the north shore of Massachusetts, the bridge has a total length of 80 feet 6 inches and is composed of two elliptical arches with a 30-foot span and 9-foot rise. In 1838, the bridge was widened 15 feet to the east or downstream side.

Choate Bridge

(1764, 1838)
US Route 1A over Ipswich River,
Ipswich, Massachusetts.
John Choate, Builder.

*A*n important and integral part of the National Road, this 80-foot span, located just west of Grantsville, a small town in western Maryland, was the largest stone arch in the country when completed. It has been bypassed by modern Route 40. The bridge's setting, the material qualities of the stonework, proportions, and detail are extraordinary.

Casselman Bridge

(1813)
Old US Route 40 over Casselman River,
Grantsville, Maryland.
David Shriver Jr., Builder.

$\mathcal{M}$asonry skills were not exclusive to the North, as this lovely Gothic arch example shows. Located in the up-country of South Carolina on the road linking Charleston and Greenville, near the small town of Tigerville, the bridge was built by the South Carolina Board of Public Works.

Poinsett Bridge

(1820)
South Carolina Road 42 over Little Gap Creek,
Tigerville vicinity, South Carolina.
Joel Poinsett, Builder.

The Washington Branch connected Washington, D.C., with the main line of the B&O at Relay, a junction southwest of Baltimore. The viaduct, built on a four-degree curve, has eight arches for a total length of 612 feet and stands 62 feet above the Patapsco River.

Thomas Viaduct

(1835)
Baltimore & Ohio Railroad, Washington Branch over Patapsco River, Elkridge, Maryland.
Benjamin Henry Latrobe Jr., Engineer.

11

$\mathscr{T}$he Canton Viaduct served the Boston and Providence Railroad. It is a granite structure 615 feet long, 22 feet wide, and 70 feet above the river. The stone walls that look like infill inside the arches are not a later addition. The structure is composed of two parallel walls, five feet thick, separated by nine feet of air space. The pilasters and arches strengthen the walls that carry the deck. The arches, however, have been reinforced with concrete as seen in the photograph. Smaller arches at the base allow for the flow of the river.

(1835)
Boston & Providence Railroad over East Branch, Neponset River, Canton, Massachusetts. Captain William Gibbs McNeil, Engineer.

$\mathscr{Canton\ Viaduct}$

$\mathcal{D}$unlaps Creek is the first all-iron bridge constructed in the United States. Designed for the National Road, the bridge has lasted all these years because the elliptical arches are made of iron castings highly resistant to corrosion. The bridge, though widened, has not been extensively modified other than the cantilevered sidewalks. It has been recognized as an engineering landmark and consequently protected since the early 20th century.

(1839)
Old US Route 40 over Dunlaps Creek,
Brownsville, Pennsylvania.
Captain Richard Delafield, Engineer.

Dunlaps Creek Bridge

$\mathcal{C}$anals preceded railroads as major developers of bridge structures. A major impediment to efficient canal operation was river crossings. Aqueducts afforded the ultimate solution. The Schoharie Creek crossing is an example of this kind of improvement, a permanent structure constructed in stone and wood sixteen years after the Erie Canal opened in 1825. The towpath ran along the arches.

Schoharie Creek Aqueduct

(1841)
Erie Canal (enlarged) over Schoharie Creek,
Fort Hunter, New York.
John B. Jervis, Engineer.

reatened by the possibility of the New York & Erie Railroad taking away its business, canal interests were powerful enough to get the New York legislature to dictate the route of the Erie Railroad through difficult terrain in the southern tier of New York State along the Pennsylvania border. One major obstacle was the broad valley of Starrucca Creek in northeastern Pennsylvania. Addressing this challenge was the construction of the Starrucca Viaduct, made up of seventeen arches, 51 feet each, for a total length of 1,040 feet and height of 110 feet above the valley floor. Construction was an equal challenge, a feat performed by Kirkwood, who was told to complete the bridge in two years no matter what the cost. Only the deep pockets of Erie Railroad financiers, a result of heavy British investment, enabled the viaduct to be built under these constraints and earned it the distinction of being the most expensive railroad structure in the world, at $320,000 in 1848 dollars.

Starrucca Viaduct

(1848)
New York & Erie Railroad over Starrucca Creek,
Lanesboro, Pennsylvania.
James P. Kirkwood and Julius W. Adams, Engineers.

Central Of Georgia Railway: Brick Arch Viaducts

*T*he engineers and architects for the Central of Georgia selected brick for the two approach viaducts leading to the railroad's terminal station and shops in Savannah, because stone was not available in coastal Georgia and brick was less expensive. Lack of stone did not denigrate the results, as seen in this detail of the little niche in the brick pilaster of the 1853 viaduct. The second viaduct, built in 1860, is shown in the drawing.

(1853, 1860)
Central of Georgia (abandoned) over
Ogeechee Canal and West Boundary Street,
Savannah, Georgia.
Augustus Schwaab, Engineer;
Martin P. Mueller, Architect.

$\mathcal{A}$ 220-foot span, this bridge stood for 39 years as the longest masonry arch in the world. It was designed for the Washington Aqueduct system, bringing water from Great Falls on the Potomac 12 miles to the distribution reservoir at Georgetown. Today, a four-lane highway passes under the arch, which is a segment of 110 degrees, a rise of 57 feet, with granite voussoirs 4 feet deep at the crown, 6 feet at the spring lines. The spandrels are locally quarried Seneca sandstone and the backing behind the arch ring is laid with radial joints, adding greatly to the strength.

Cabin John Aqueduct

(1864)
Washington Aqueduct over Cabin John Creek,
Cabin John, Maryland.
Montgomery C. Meigs, Engineer.

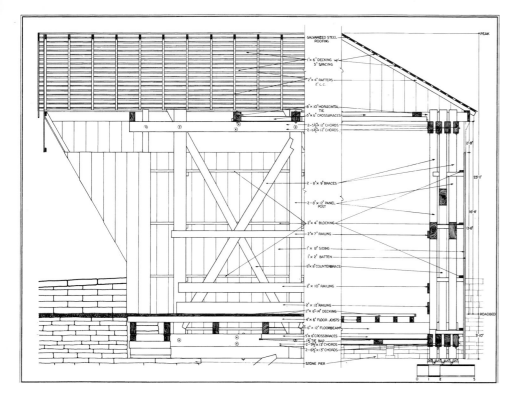

The photograph shows the trusses exposed while the bridge was being dismantled prior to its relocation and storage in Indianapolis in 1976. Ten years later, the bridge was moved again and restored to pristine condition in Mill Race Park, Columbus, Indiana.

The photograph and drawing compare truss framing details. Brownsville Bridge is designated in the *World Guide to Covered Bridges* as No. 140304 and is a fine example of a Long truss. Ninety-one covered bridges remain in Indiana.

Brownsville Bridge

(1840)
Originally Main Street over Whitewater River, Brownsville, Indiana.
Builder unknown.

This structure is the oldest all-metal railroad bridge still in active use in the United States, though that use is vehicular and not rail. The bridge provides access to a farmer's property across former Reading Railroad tracks. It is also second of the three oldest surviving all-metal railroad bridges in the United States. The first is the Manayunk Bridge (1845), designed by the same engineer for the same railroad. A single surviving truss from the former bridge is presently displayed at the Smithsonian Institution as part of the "John Bull" locomotive exhibit. The third is the Haupt Truss (c.1854), now disassembled at the Railroaders' Memorial Museum in Altoona, Pennsylvania. As originally erected on the Reading main line, this bridge was a 69-foot, 18-panel Howe truss with cast iron diagonal compression members, wrought iron vertical tension rods and wrought iron bars for the top and bottom chords. Halls Station Bridge has Egyptian Revival motifs in its castings.

Reading–Halls Station Bridge

(c.1846)
Private road over Conrail,
Muncy, Pennsylvania.
Richard B. Osborne, Engineer.

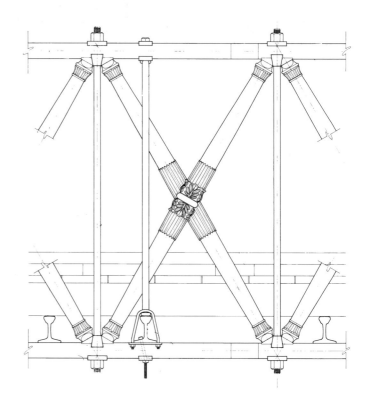

TYPICAL PANEL

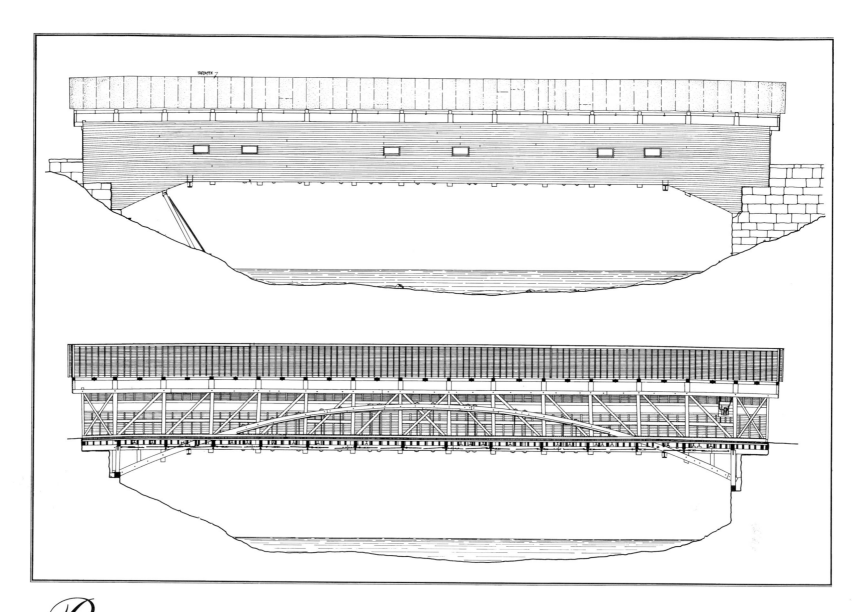

$\mathcal{B}$arrickville, one of the few two-lane covered bridges, is a classic example of a Burr arch-truss, as shown in the drawings.

Barrickville Covered Bridge

(1853)

Buffalo Creek, Barrickville, West Virginia.

Lemuel Chenoweth, Builder.

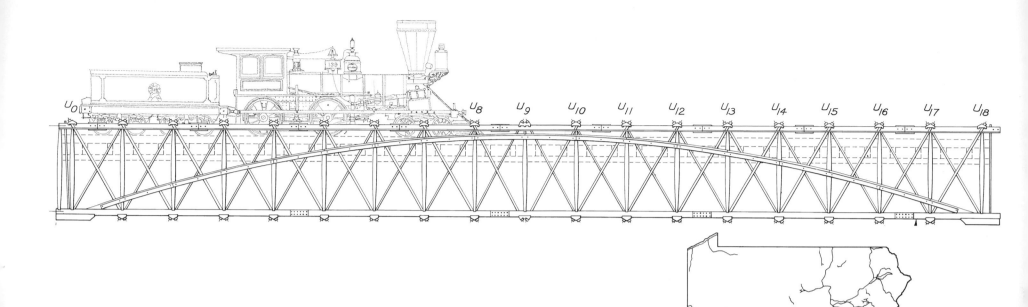

This bridge is similar to the Burr arch-truss, a wooden trussing system in which an arch is superimposed on a rectangular truss to increase its strength. The Haupt bridge is a combined Pratt truss and tied arch made of iron. Herman Haupt was European-born, but studied engineering at West Point. He was chief engineer of the Pennsylvania Railroad from 1848 to 1856. The first Haupt truss was built at the Pennsylvania's Altoona shops in 1851 and was the railroad's first all-iron bridge. Originally located at Vandevander, Pennsylvania, this bridge was moved to Thompsontown in 1889, where it remained in service as a vehicular bridge over the main line until 1984, when it was moved to the railroad museum in Altoona. Three other Haupt truss bridges survive, at Ronks, Ardmore, and Villanova. All are single spans over the main line tracks.

State Railroad Map ca. 1860

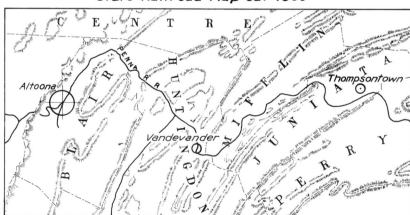

Haupt Truss Bridge

(c.1854)
Railroaders' Memorial Museum,
Altoona, Pennsylvania.
Herman Haupt, Engineer.

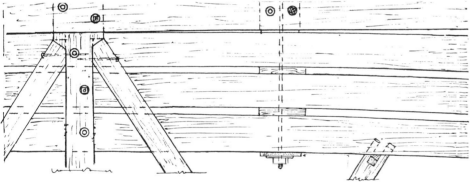

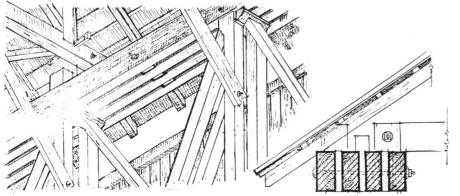

*B*lenheim Bridge, in addition to being a rare example of a "double-barreled" bridge using three lines of trussing, is the second-longest single-span covered bridge in the world after the Bridgeport Bridge in California. Powers selected Long trusses, patterned after a design patented by Colonel Stephen H. Long in 1830, because the Long truss was capable of spanning the 210 feet required at this crossing. The densely composed drawing sheet of the bridge gives details of the truss framing.

Blenheim Bridge

(1855)
Schoharie Creek, North Blenheim, New York.
Nicholas Montgomery Powers, Builder.

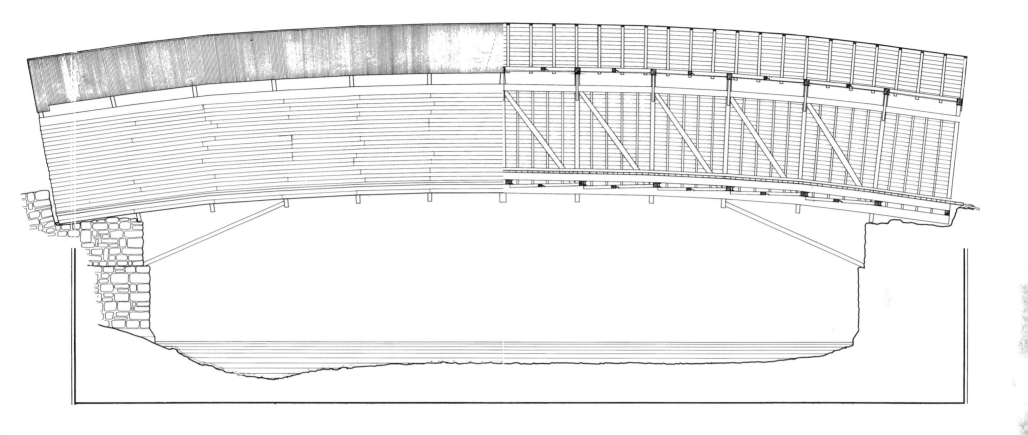

Humpback Covered Bridge

The eight-foot rise that gives the Humpback Bridge its distinctive appearance is unique in the United States. The curve is built into the top and bottom chords of the multiple king-post trussing system. The bridge was built by the James River and Kanawha Turnpike; there is no record of its designer. Legend has it that Northern and Southern troops negotiated an agreement that prevented the bridge from being destroyed during the Civil War.

(1857)
Dunlaps Creek, Covington vicinity, Virginia.
Builders unknown.

Bridgeport Covered Bridge

*T*his is the longest single-span covered bridge in the world, built for the Virginia City Turnpike Company, which linked the port city of San Francisco, the source of manpower and supplies, with the Comstock Lode in western Nevada, the world's greatest source of silver in the late 19th century. The trace of the wooden arch is reflected on the exterior shingle siding, a fine architectural detail. The interior view shows the arch superimposed on the Howe truss to carry the unprecedented 233-foot span.

(1862)
South Fork Yuba River, Grass Valley, California.
David Wood, Builder.

Cornish–Windsor Covered Bridge

(1866)

State Route 44 over Connecticut River,
Cornish, New Hampshire–Windsor, Vermont.
James Tasker and Bela Fletcher, Builders.

Debate swirled for years about the most appropriate way to restore this bridge. One group advocated a "traditional" solution of adding arches. The other claimed this would destroy the character-defining elements of the bridge and was bad engineering. Eventually, it was decided to maintain the original appearance of the bridge not by adding any new structure but rather by strengthening the original Town lattice trusses by splicing glue-laminated beams onto the top and bottom chord members. Glue-laminated beams, perfected in the 1930s, provide great strength for a small cross-sectional area and made it possible to "needle" the beams into the existing structure and attach them unobtrusively. While needling the beams into place, the bridge was temporarily supported by a cable-stayed suspension system that allowed the engineers to eliminate the downstream sag that had crept into the structure over the years and to restore the original camber (slight arching) to the trusses. This innovative engineering solution has resulted in a bridge capable of carrying school buses and emergency equipment while at the same time preserving one of New England's premiere engineering monuments.

Several engineering historians consider the Delaware Aqueduct equal in significance to the Brooklyn Bridge because it is the oldest suspension bridge in America. Its designer, John Roebling, continued the thinking initiated by Finley, and with this structure, perfected suspension bridge technology. The bridge actually was an aqueduct that carried the Delaware & Hudson Canal across the Delaware River. Abandoned in 1898, the aqueduct was converted to highway use. In 1980, it was purchased by the National Park Service as part of the newly established Upper Delaware Scenic & Recreational River. Over the past few years, the Park Service has rehabilitated the aqueduct and reconstructed the superstructure to its original appearance. This work won a Presidential Design Award for excellence in 1988. The isometric view shows all the principal elements of this aqueduct structure at the abutment.

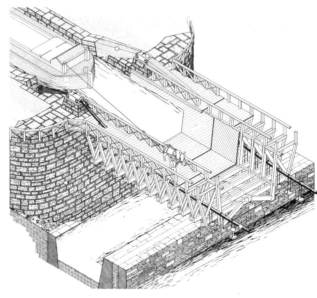

Delaware Aqueduct

(1849)
Delaware River, Lackawaxen, Pennsylvania.
Minisink Ford, New York.
John Roebling, Engineer.

$\mathcal{C}$ompleted months after Roebling's Delaware Aqueduct, Ellet's bridge is no less significant. It had the distinction of being the longest span in the world with a clear span between towers of 1,010 feet. The major technological feature of both these early bridges is the use of individual wires, bundled into cables, as seen in this detail of the suspension tower. The design of the cables on the Wheeling bridge was originally based on the French system. Rather than a single cable, Ellet used "garlands," or multiple cables, to carry the deck. It was not until later, after several of the cables were blown off the towers during a storm, that they were bound into a single cable. The bundled cable gives the suspension bridge its strength and the potential of such tremendous spans. The longest bridges to this day are suspension bridges.

Wheeling Suspension Bridge

(1849)
Old Route 40 over Ohio River, Wheeling, West Virginia.
Charles Ellet Jr., Engineer.

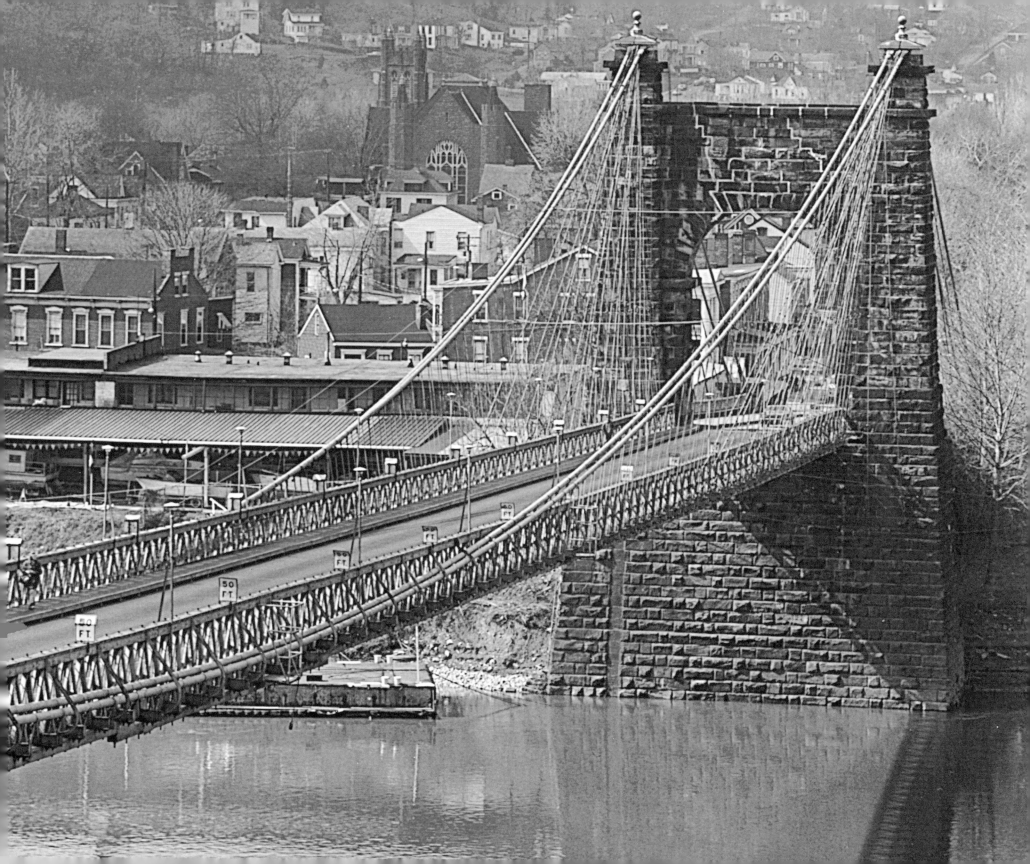

$\mathcal{I}$n 1866, this was the longest spanning bridge in America and the longest span in the world, at 1,057 feet. Designed by John Roebling, cable spinning and final construction was supervised by his son, Washington. This project nurtured the father-son relationship that proved critical to completion of the Brooklyn Bridge 16 years later. Because of the importance of the Ohio crossing at Cincinnati, the bridge was one of few private civil works projects allowed to continue during the Civil War. By 1895 loads had increased to such a degree that new steel stiffening trusses were added. In 1899, additional cables were added and squat sheet-metal domes replaced Roebling's original finials on the towers. The original finials have been restored since these photographs were taken.

Cincinnati Suspension Bridge

(1866)
Ohio River, Cincinnati, Ohio-Covington, Kentucky.
John Roebling, Engineer.

This bridge is a lovely vernacular example displaying the quintessence of straightforward Yankee ingenuity. The anchorages, where the cables are splayed into their individual strands and change direction over the large stones, clearly show the same principle used in longer and more sophisticated suspension bridges. The shingle-sided towers were rehabilitated by the Maine Highway Department in the 1960s. The designer of this 198-foot span and the cable maker are not known.

New Portland Wire Bridge

(c.1868)
Carrabasset River, New Portland, Maine.
Builder unknown.

Part Two

TIME LINE 1828–1850

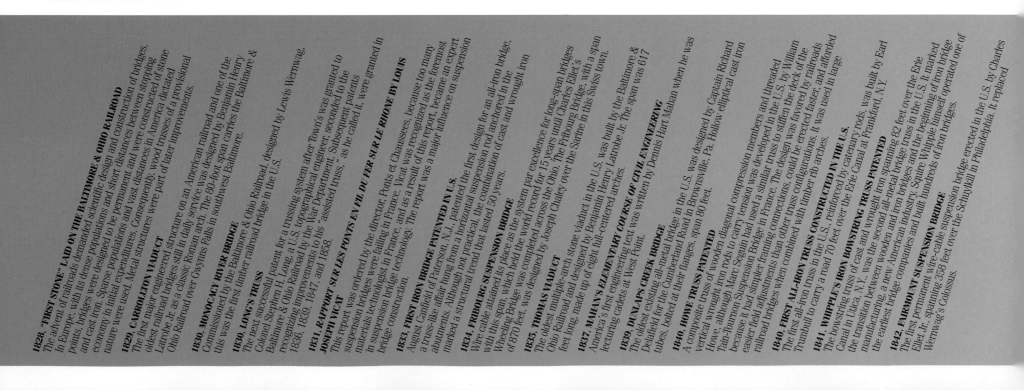

1828, "FIRST STONE" LAID ON THE BALTIMORE & OHIO RAILROAD
The advent of railroads demanded scientific design and construction of bridges. In Europe, with its dense populations and short distances between stopping points, bridges were designed to be permanent and were constructed of stone and cast iron. Sparse populations in America dictated economy in initial expenditures. Metal structures. Consequently, wood trusses were used.

1829, CARROLLTON VIADUCT
The first major engineered structure on an American railroad and one of the oldest railroad bridges still in daily service. The 80-foot span was designed by Benjamin Henry Latrobe Jr. as a classic Roman arch.

1830, MONOCACY RIVER BRIDGE
Commissioned by the Baltimore & Ohio Railroad, designed by Benjamin Henry Latrobe. this was the first timber railroad bridge in southwest Baltimore.

1830, LONG'S TRUSS
The next successful patent for a trussing system after Town's was granted to Colonel Stephen H. Long, a U.S. topographical engineer. Subsequent patents recognizing improvements to his "assisted truss," as he called it, were granted in 1836, 1839, 1847, and 1858.

1831, RAPPORT SUR LES PONTS EN FIL DU FER SUR LE RHONE BY LOUIS JOSEPH VICAT
This report was ordered by the director, Ponts et Chaussees, because too many suspension bridges were failing in France. Vicat was recognized as the foremost materials technologist in France, and as a result of this report, became an expert in suspension bridge technology. The report was a major influence on suspension bridge construction.

1833, FIRST IRON BRIDGE PATENTED IN U.S.
August Canfield of Paterson, N.J., patented the first design for an all-iron bridge, a truss-like affair hung from a horizontal suspension rod anchored in the abutments. Although not practical, the combination of cast and wrought iron marked a structural trend that lasted 50 years.

1834, FRIBOURG SUSPENSION BRIDGE
Wire cable attained its place as the system par excellence for long-span bridges with this span, which held the world record for 15 years until Charles Ellet's Wheeling Bridge was completed across the Ohio. The Fribourg Bridge, with a span of 870 feet, was designed by Joseph Chaley over the Sarine in this Swiss town.

1835, THOMAS VIADUCT
The oldest multiple-arch stone viaduct in the U.S. was built by the Baltimore & Ohio Railroad, and designed by Benjamin Henry Latrobe Jr. The span was 617 feet long, made up of eight full-centered arches.

1837, MAHAN'S ELEMENTARY COURSE OF CIVIL ENGINEERING
America's first engineering text was written by Dennis Hart Mahan when he was lecturing cadets at West Point.

1839, DUNLAPS CREEK BRIDGE
The oldest existing all-metal bridge in the U.S. was designed by Captain Richard Delafield for the Cumberland Road in Brownsville, Pa. Hollow elliptical cast iron tubes, bolted at their flanges, span 80 feet.

1840, HOWE TRUSS PATENTED
A composite truss of wooden diagonal compression members and threaded vertical wrought iron rods to carry tension was developed in the U.S. by William Howe, although Marc Seguin had used a similar truss to stiffen the deck of the Tain-Tournon Suspension Bridge in France. The design was favored by railroads because it had simpler framing connections, could be erected faster, and afforded easier field adjustments than other truss configurations. It was used in large railroad bridges when combined with timber rib arches.

1840, FIRST ALL-IRON TRUSS CONSTRUCTED IN THE U.S.
The first all-iron truss in the U.S., reinforced by catenary rods, was built by Earl Trumbull to carry a road 70 feet over the Erie Canal at Frankford, N.Y.

1841, WHIPPLE'S IRON BOWSTRING TRUSS PATENTED
The bowstring truss of cast and wrought iron spanning 82 feet over the Erie Canal in Utica, N.Y., was the second all-metal iron bridge in the U.S. It marked the transition between iron and wooden bridges and the beginning of iron bridge manufacturing. a new American industry. Squire Whipple himself operated one of the earliest bridge companies and built hundreds of iron bridges.

1842, FAIRMOUNT SUSPENSION BRIDGE
The first permanent wire-cable suspension bridge erected in the U.S. by Charles Ellet Jr., spanning 358 feet over the Schuylkill in Philadelphia. It replaced Wernwag's Colossus.

Civil War Era Metal Truss Bridges

The next group of bridges contains the most extraordinary examples in the United States. Some have survived for nearly 140 years and represent a category that shows the change in bridge building technology from carpenter-craftsmen to engineered mass produced prefabrication. The change in materials from wood and stone to cast and wrought iron and eventually, in 1890, to steel, engendered great strength and longer spans. Many of these transitional bridges were destroyed during the Civil War, others were lost during scrap drives of the two World Wars. Most were replaced as greater vehicular weights and speeds made them obsolete. Other reasons that so few remain are that cast and wrought iron are less durable than stone; they are not recognized by the public, preservationists, or the engineering profession as being as significant as the more monumental Brooklyn or Golden Gate suspension bridges, or the quaint and nostalgic stone arches or wooden covered spans. Though hundreds of these bridges were built from 1840 to 1880, only a handful survive.

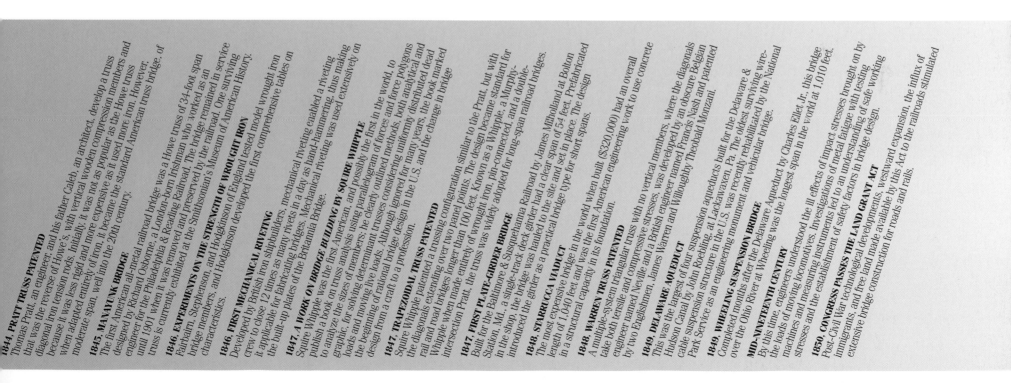

1844, PRATT TRUSS PATENTED
Thomas Pratt, an engineer, and his father Caleb, an architect, develop a truss that was the reverse of Howe's, with vertical wooden compression members and diagonal iron tension rods. Initially, it was not as popular as the Howe truss because it was less rigid and more expensive as it used more iron. However, when adapted entirely of iron, it became the standard American truss bridge, of moderate span, well into the 20th century.

1845, MANAYUNK BRIDGE
The first American all-metal railroad bridge was a Howe truss of 34-foot span designed by Richard Osborne, a London-born Irishman who worked as an engineer for the Philadelphia & Reading Railroad. The bridge remained in service until 1901 when it was removed and preserved by the railroad. One surviving truss is currently exhibited at the Smithsonian's Museum of American History.

1846, EXPERIMENTS ON THE STRENGTH OF WROUGHT IRON
Developed by British iron shipbuilders, Fairbairn, Stephenson, and Hodgkinson developed the first comprehensive tables on bridge members, and Hodgkinson tested model wrought iron characteristics.

1846, FIRST MECHANICAL RIVETING
Developed by British iron shipbuilders, mechanical riveting enabled a riveting crew to close 12 times as many rivets in a day as hand-hammering, thus making it applicable for fabricating bridges. Mechanical riveting was used extensively on the built-up plates of the Britannia Bridge.

1847, A WORK ON BRIDGE BUILDING BY SQUIRE WHIPPLE
Squire Whipple was the first American, and possibly the first in the world, to publish a book on truss analysis using parallelogram of forces and force polygons graphic, for solving the sizes of members; he clearly outlined methods, both analytical and graphic, for solving determinant trusses and moving live loads. With this it was the beginning of rational trusses considering uniformly distributed dead loads and moving live loads. Although ignored for many years, the book marked the beginning of rational bridge design from a craft to a profession.

1847, TRAPEZOIDAL TRUSS PATENTED
Squire Whipple patented a trussing configuration similar to the Pratt, but with the diagonals extending over two panel points. The design became standard for rail and road bridges longer than 100 feet. Known as a Whipple, a Murphy-Whipple when made entirely of wrought iron, pin-connected, and a double-intersection Pratt, the truss was widely adopted for long-span railroad bridges.

1847, FIRST PLATE-GIRDER BRIDGE
Built for the Baltimore & Susquehanna Railroad by James Milholland at Bolton Station, Md., a single-track deck girder had a clear span of 54 feet. Prefabricated in the shop, the bridge was hauled to the site and set in place. The design introduced the girder as a practical bridge type for short spans.

1848, STARRUCCA VIADUCT
The most expensive bridge in the world when built ($320,000) had an overall length of 1,040 feet and was the first American engineering work to use concrete in a structural capacity in its foundation.

1848, WARREN TRUSS PATENTED
A multiple-system triangular truss with no vertical members, where the diagonals take both tensile and compressive stresses, was developed by an obscure Belgian engineer named Neuville and a British engineer named Francis Nash and patented by two Englishmen, James Warren and Willoughby Theobald Monzani.

1849, DELAWARE AQUEDUCT
This was the largest of four suspension aqueducts built for the Delaware & Hudson Canal by John Roebling, at Lackawaxen, Pa. The oldest surviving wire-cable suspension structure in the U.S. was recently rehabilitated by the National Park Service as an engineering monument and vehicular bridge.

1849, WHEELING SUSPENSION BRIDGE
Completed months after the Delaware Aqueduct by Charles Ellet, Jr., this bridge over the Ohio River at Wheeling was the longest span in the world at 1,010 feet.

MID-NINETEENTH CENTURY
By this time, engineers understood the ill effects of impact stresses brought on by the loads of moving locomotives. Investigations of metal fatigue with testing machines and measuring instruments led to an understanding of safe working stresses and the establishment of safety factors in bridge design.

1850, CONGRESS PASSES THE LAND GRANT ACT
Post-Civil War technological developments, westward expansion, the influx of immigrants, and free land made available by this Act to the railroads stimulated extensive bridge construction for roads and rails.

Fink Through-Truss Bridge

For many years this was among the oldest surviving metal trusses in the United States and the best example of Albert Fink's 1854 patent. Cast and wrought iron trusses have no built-in redundancy. When the cast iron end post on this bridge was hit by an automobile in 1978, the bridge collapsed. Although no one was injured, America lost an engineering landmark due to the lack of an inexpensive guardrail. Albert Fink, like John Roebling, Herman Haupt, and others, was one of the immigrant engineers who through their superior European training, helped advance engineering practice in the United States. Hunterdon County, where the bridge was located, was one of the first counties to select an all-metal bridge rather than a wooden or stone one.

(1858, destroyed 1978)
South Branch Raritan River,
Hamden, New Jersey.
Trenton Locomotive & Machine Works, Fabricator.

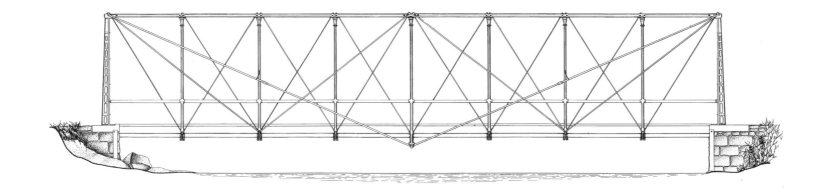

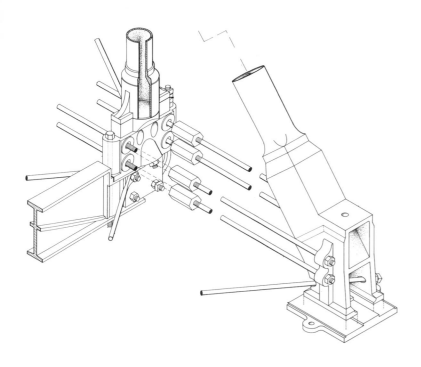

Walnut Street Bridge

B eckel fabricated this bridge in his family's foundry, the Beckel Iron Foundry & Machine Shops, in nearby Bethlehem. Though not a graduate engineer, Beckel took bridge-engineering courses under Francis Lowthorp at Lehigh University and used Lowthorp's patented details in this structure. It is said that God is in the details, and it is the details that are the most interesting features of cast and wrought iron bridges, particularly bridges based on the Lowthorp patent. The integral cast iron joint block and floor beam is designed so that the bottom-chord rods pass through with no rigid connection. This detail avoided what engineers call a "stress" at the junction of the vertical and horizontal connections. The drawing also shows the end of the cast iron floor beams. Although cast iron is rarely used in beams, Beckel designed cast iron beams with refinements that successfully withstood loads without failure for more than 90 years.

(1860)
Formerly spanning Saucon Creek,
Hellertown, Pennsylvania.
Charles N. Beckel, Builder.

45

Bow Bridge

The most familiar of the Central Park bridges because of its location in a popular section of the park, Bow Bridge has the longest span at 87 feet. Wrought iron truss rods that strengthen the transverse deck beams can be seen on the underside of the deck. Bow Bridge has the further distinction of being the oldest wrought iron girder bridge surviving in the United States. This was revealed in 1971 when the decorative castings were removed and the wrought iron plate girders, slightly arched and pierced where the cinquefoils appear in the balustrade, were exposed during restoration.

(1862)

The Lake, Central Park, New York, New York. Calvert Vaux and Jacob Wrey Mould, Designers.

Pine Bank Arch

This is the first of five remarkable cast iron arch bridges that survive in Central Park. Together, they constitute the oldest cast iron bridges in the United States (other than Dunlaps Creek) and with the other Central Park bridges, the greatest collection of decorative park bridges in the country. Restoration of Pine Bank Arch in 1984 included recasting missing parts, replacing the deck, painting, and planting a grove of white pines on the eastern embankment.

(1861)

Central Park, New York, New York. Calvert Vaux and Jacob Wrey Mould, Designers.

Reservoir Bridge Southeast

(1865)
Central Park, New York, New York.
Calvert Vaux and Jacob Wrey Mould, Designers.

48

Delicate decorative castings distinguish this 33-foot span iron arch bridge.

Reservoir Bridge Southwest

*A*nother of the cast iron arches separating pedestrian from equestrian traffic, this bridge spans 72 feet.

(1864)
Central Park, New York, New York.
Calvert Vaux and Jacob Wrey Mould, Designers.

Bridge 28 (Gothic Arch)

*B*ridge 28 is not only one of the earliest cast iron arches in the United States, but one of the few constructed. Cast iron arches were standard for European spans in the late 18th and the first half of the 19th century, but were not common in the United States. The arch is 37 feet 5 inches long and 15 feet 3 inches high. Calvert Vaux and Jacob Wrey Mould worked with landscape architect Frederick Law Olmsted. The Gothic motifs with sinuous art nouveau lines actually predate this design style, which did not appear in Europe until 20 years later.

(1864)
Central Park, New York, New York.
Calvert Vaux and Jacob Wrey Mould, Designers.
J.B. & W.W. Cornell Ironworks of New York, Fabricators.

Upper Pacific Mills Bridge

One of the rarest and most significant early iron bridges was nearly destroyed but for a fortuitous Sunday morning drive by Francis Griggs, a professor of civil engineering at nearby Merrimack College. He and his wife came upon the bridge while a salvage company was cutting it up for scrap. Professor Griggs was able to negotiate with the owner for removal of the bridge to Merrimack College, where engineering students under his direction restored the bridge as a class project. The bridge was designed and patented by Thomas Moseley, who introduced the riveted wrought iron bridge to America. Wrought iron bridges became standard 20 years later. The most interesting feature of the bridge, other than the triangular upper chord, are the two counterbraces seen as reverse curves to the main arch. These prevented deflection of that arch under dynamic live loads such as a heavy wagon being pulled across the bridge.

(1864)
North Canal, Lawrence, Massachusetts.
Thomas Moseley, Designer.

These photographs capture the essence of Squire Whipple's contribution to bridge technology: a spidery structure of iron, easily erected, using repetitive parts of minimal material, that successfully defied gravity. Whipple patented the bowstring truss in 1841 and built the first one over the Erie Canal. He formalized his theories in *A Work on Bridge Building*, published in 1847, the first American text on truss analysis using parallelogram of forces and force polygons to analyze stresses and thus determine the size of connections and members exactly. Whipple's book and innovations mark a fundamental change in bridge design from a craft tradition to an engineering profession: the transition between bridges built of stone and wood to iron and the beginning of a new American industry iron bridge manufacturing. Squire Whipple was one of the first entrepreneurs to start a company exclusively engaged in building metal structures.

Whipple Cast- & Wrought Iron Bowstring Truss Bridge

(1869)
Normans Kill Private Way near Normans Kill Ravine, Albany, New York.
Squire Whipple, Engineer,
Simon DeGraff, Builder

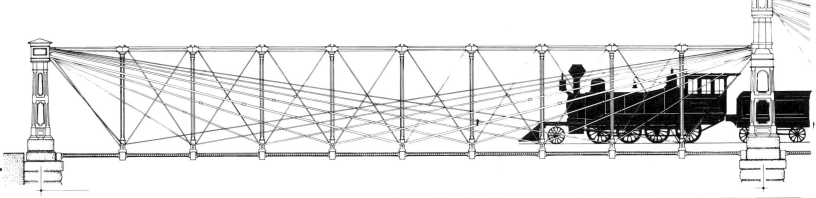

This bridge is to America what Ironbridge is to England, not the oldest but the most recognizable monument to iron bridge technology in the United States, though few people in America are aware of the Bollman Truss Bridge at Savage. The picture shows the intricate castings of the upper end posts where the wrought iron eye-bars are pinned into place. The bridge has since been restored, the distinctive wooden housings protecting the upper end-post connections replicated, and the superstructure painted iron-oxide red as part of the Little Patuxent River Park at the Savage Mill. The drawing of a Bollman bridge that crossed the Potomac at Harpers Ferry gives the overall configuration of the Bollman design's composite cast and wrought iron construction consisting of a pair of wrought iron diagonal eye-bars extending from the ends of the top chord down to each panel point.

Bollman Truss Bridge

(1869)
Little Patuxent River, Savage, Maryland.
Wendel Bollman, Engineer.

First appearances of this bridge are deceiving; it looks much more modern than its date. When documented by HAER during the summer of 1991, it was discovered that the bridge is a Moseley design, based on his 1866 patent for a "Wrought Iron Arch Girder Bridge." Moseley is known primarily for his riveted tubular arches fabricated in boilerplate like the Upper Pacific Mills Bridge. But research conducted in 1991, as part of a HAER and West Virginia University project to document the surviving cast and wrought iron bridges in America, confirmed that Moseley had an iron bridge company in Philadelphia where this bridge was fabricated. Characterized by the arched upper chord formed by a pair of Z-bars riveted at their flanges and the lattice webbing, this 103-foot span is the only surviving example of Moseley's wrought iron lattice girder bridge.

Hare's Hill Road Bridge

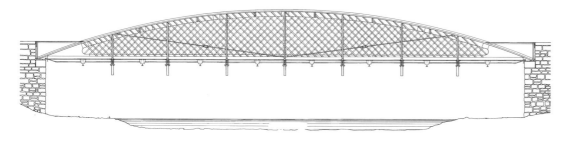

(1869)
Hare's Hill Road over French Creek,
Kimberton, Pennsylvania.
Thomas Moseley, Builder.

54

Another one-and-only known survivor of the patented prefabricated metal-truss designs that competed in the bridge market in the years following the Civil War, this one was patented in 1869 by Joseph G. Henszey of Philadelphia. It was manufactured by the Continental Bridge Company, also of Philadelphia, which was in business from 1869 to 1878. That the design resembles Moseley's tubular tied-arch bridge is no coincidence, as he had recently opened an office in Philadelphia and was marketing his bridges in the surrounding counties. Henszey's bridge is different in the make-up of the upper chord member and the trussing arrangement at the lower chord level, but the vertical straps in the web of the truss and the nearly all-riveted fastenings are similar to Moseley's design.

Henszey's Wrought Iron Arch Bridge

(1869)
Kings Road over Ontelaunee Creek,
Wanamakers, Pennsylvania.
Joseph G. Henszey, Designer.

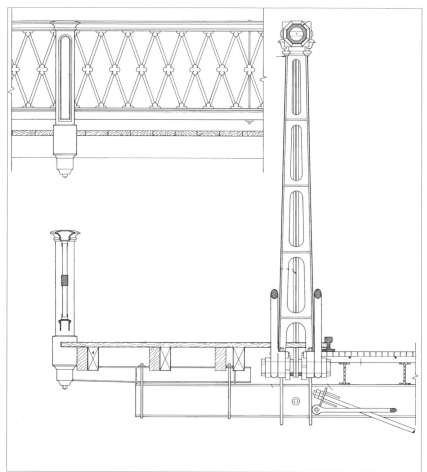

This is one of three similar bridges surviving in New Jersey manufactured by William and Charles Cowin, owners of the Lambertville Iron Works in Lambertville, New Jersey; the others are located in little milling villages along the Musconetcong River in New Hampton (1868) and Spruce Run in Glen Gardener (1870). All are composite cast and wrought iron Pratt pony trusses based on the patents of Francis C. Lowthorp, a well-known engineer of Trenton, New Jersey. The Clinton bridge incorporates another interesting feature, a tension adjuster patented by William Johnson in 1870. The drawing shows the intricate details of the castings for the vertical posts, the octagonal upper chord member, and the railing that attests to the skill of the pattern maker as much as the engineer. The deck beams and wearing surface of the deck are the only major elements of the bridge that have been replaced in 120 years.

West Main Street Bridge

(1870)
West Main Street over South Branch Raritan River, Clinton, New Jersey.
William and Charles Cowin, Builders.

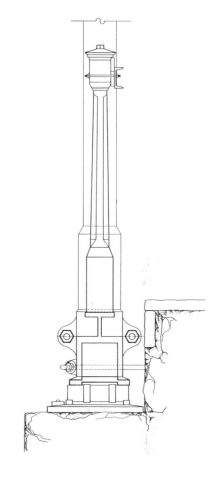

Although this bridge was built by a competitor of Charles and William Cowin of the Lambertville Iron Works, it too includes bridge details patented by Francis C. Lowthorp. The bridge was manufactured by the same iron-works that built the Walnut Street Bridge just a few miles upstream. Though Lowthorp was the inspiration behind the bridges, equal credit must go to Charles Beckel, master foundryman, whose studies of bridge design with Lowthorp brought the best of two disciplines—craftsman-ship and technical design theory—to bridge design. Like the Walnut Street Bridge, the Old Mill Road Bridge incorporates the unique cast iron beam that is integral with the bottom chord connection. Evidence of the soundness of the design is the fact that the deck beam was not reinforced until the 1930s, a sixty-year period during which loads on the bridge changed dramatically.

Old Mill Road Bridge

(1870)
Old Mill Road over Saucon Creek,
Hellertown, Pennsylvania.
Charles N. Beckel, Builder.

This bridge was built by the Keystone Bridge Company of Pittsburgh, a company owned by Andrew Carnegie that eventually became US Steel. During the 1870s and '80s, it was a major manufacturer of iron railroad bridges under the leadership of Jacob H. Linville, bridge engineer. Keystone bridges are recognized by the distinctive split compression members.

Several Keystone bridges survive around the country; the one on the left located at the time of the photograph near Kassler, Colorado, spanning the South Fork of the South Platte River. It originally served the Denver, South Park & Pacific Railroad before being relocated for vehicular service in 1935. In 1978, the bridge was match-marked and reassembled at another location. Judging by the more decorative detailing of the end-post castings and the bosses on the column spacers, this bridge may predate 1870, perhaps as early as 1865.

Stewartstown Railroad Bridge

(1870)
Stewartstown Railroad over Valley Road,
Stewartstown, Pennsylvania.
Jacob H. Linville, Engineer.

Elm Street Bridge

The 110-foot span fabricated by National Bridge & Iron Works of Boston is based on a patent granted to Charles H. Parker, company engineer, in 1870. By c.1910, loads had increased to the point that Warren trusses were added under the deck to increase the carrying capacity of the Parker trusses. In 1975, the bridge was rated deficient and scheduled for replacement. Negotiation between the town and highway officials resulted in a precedent-setting compromise in which federal highway funds were used to rehabilitate a bridge that did not meet AASHTO standards. The waiver was granted because of the bridge's historical significance.

(1870)
Elm Street over Ottauquechee River,
Woodstock, Vermont.
Charles H. Parker, Engineer.

Riverside Avenue Bridge

(1871)
Riverside Avenue over Amtrak,
Greenwich, Connecticut.
Francis C. Lowthorp, Engineer.

This lovely specimen of the founder's art recalls a time when bridges were advertisements for the railroad that built them—in this example, the New York, New Haven & Hartford. This double-intersection Pratt truss originally was part of a six-span structure over the Housatonic River in Stratford, Connecticut. In the 1890s, when the bridge was replaced with a stronger structure, this single span was relocated to its present site. Aside from the decorative castings of the portal and horizontal struts in the plane of the upper chord, the main feature of the bridge is the piercing of the vertical posts by the diagonals. This detail kept the forces aligned on the centerline of the truss rather than eccentrically, had the diagonals been placed inside the vertical posts, which was common for other double-intersection Pratt trusses. The bridge was recently rehabilitated a second time to ensure its continued use and preservation as an engineering landmark.

The most graceful of the Mississippi River bridges, also the first extensive use of steel in bridge construction, shattered engineering precedents of the time. Its three arch spans of 502, 520, and 502 feet were by far the longest of its day. To erect the arches without centering, Eads used the ancient technique of cantilevering. Each span has four truss-stiffened arches with parallel chords, 12 feet apart, made of wrought iron tubes, 16 inches in diameter, with chrome steel staves inside. The depth of the river necessitated pneumatic caissons, rather than cofferdams, to build the piers.

Eads Bridge

(1874)
Over the Mississippi River at Washington Street,
St. Louis, Missouri.
James B. Eads, Engineer.

William H. Allen and Oliver H. Perry patented the design for this bridge in 1871. The company of Allen, McEvoy & Company, machinists and general jobbers, located just over the Iowa border in Beloit, Wisconsin, manufactured agricultural machinery, farm implements and this bridge. Nothing is known of Perry other than that he was a carpenter. This bridge is a classic example of the prefabricated iron-truss bridge patented and manufactured by local entrepreneurs for local farm-to-market road use. The bridge is barely a 25-foot span, but supported full milk trucks, attesting to good engineering, when these photographs were taken. The bridge has been relocated to City Park in Castalia, Iowa.

Eureka Wrought Iron Bridge

(c.1871)
Tributary of Yellow Creek, Frankville, Iowa.
William H. Allen and Oliver H. Perry,
Designers.

66

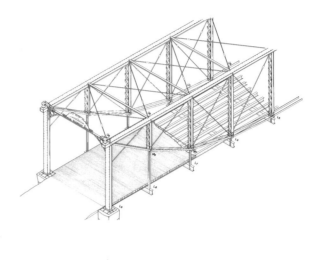

Part Three

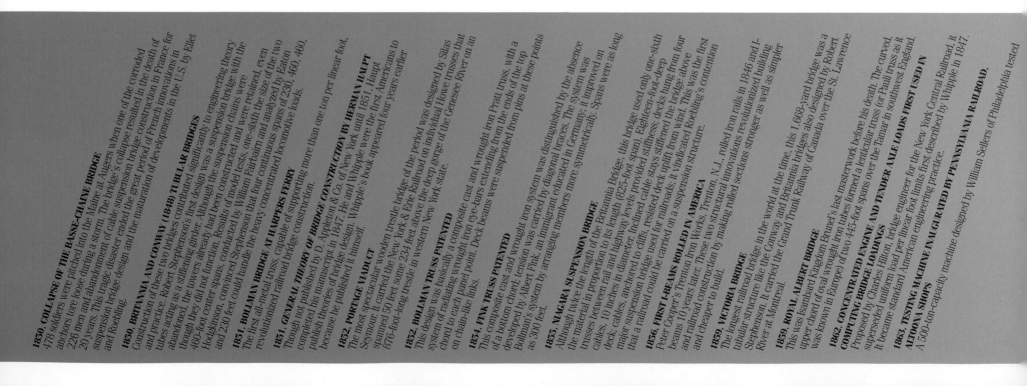

TIME LINE 1850–1875

1850, COLLAPSE OF THE BASSE-CHAINE BRIDGE
478 soldiers were pitched into the Maine at Augers when one of the corroded anchors tore loose during a storm. The bridge's collapse resulted in the death of 226 men and abandonment of cable suspension bridge construction in France for 20 years. This tragic disaster ended the great period of French innovations in suspension bridge design and the maturation of developments in the U.S. by Ellet and Roebling.

1850, BRITANNIA AND CONWAY (1848) TUBULAR BRIDGES
Construction of these two bridges contributed significantly to engineering theory and practice. Robert Stephenson's first design was a suspension bridge with the tubes acting as a stiffening girder. Although they did not function, even though the towers already had been constructed and were retained, even Hodgkinson, convinced Stephenson that four continuous spans of 230, 460, 460, 460-foot center spans, conducted by William Fairbairn and analyzed by Eaton and 230 feet could handle the heavy concentrated locomotive loads.

1851, BOLLMAN BRIDGE AT HARPERS FERRY
The first all-metal truss, capable of supporting more than one ton per linear foot, revolutionized railroad bridge construction.

1851, GENERAL THEORY OF BRIDGE CONSTRUCTION BY HERMAN HAUPT
Though not published by D. Appleton & Co. of New York until 1851, Haupt completed his manuscript in 1847. He and Whipple were the first Americans to publish theories of bridge design. Whipple's book appeared four years earlier because he published it himself.

1852, PORTAGE VIADUCT
The most spectacular wooden trestle bridge of the period was designed by Silas Seymour. It carried the New York & Erie Railroad on individual Howe trusses that spanned 50 feet some 234 feet above the deep gorge of the Genesee River on an 876-foot-long trestle in western New York state.

1852, BOLLMAN TRUSS PATENTED
This design was basically a composite cast and wrought iron Pratt truss, with a system of radiating wrought iron eye-bars extending from the ends of the top chord to each panel point. Deck beams were suspended from pins at these points on chain-like links.

1854, FINK TRUSS PATENTED
This composite cast and wrought iron system was distinguished by the absence of a bottom chord; tension was carried by diagonal braces. The system was developed by Albert Fink, an immigrant educated in Germany; it improved on Bollman's system by arranging members more symmetrically. Spans were as long as 300 feet.

1855, NIAGARA SUSPENSION BRIDGE
Although twice the length of the Britannia Bridge, this bridge used only one-sixth the material in proportion to its length (825-foot span). Eighteen-foot-deep trusses between rail and highway levels provided stiffness; decks hung from four cables, 10 inches in diameter anchored to cliffs, resisted deck uplift from wind. This was the first deck; cables, anchored to cliffs, resisted deck uplift from wind. This was the first major suspension bridge used for railroads. Inclined cable stays stiffened the bridge above that a railroad could be carried on a suspension structure. It vindicated Roebling's contention

1856, FIRST I-BEAMS ROLLED IN AMERICA
Peter Cooper's Trenton Iron Works, Trenton, N.J., rolled iron rails in 1846 and I-beams 10 years later. These two structural innovations revolutionized building and railroad construction by making rolled sections stronger as well as simpler and cheaper to build.

1859, VICTORIA BRIDGE
The longest railroad bridge in the world at the time, this 1,668-yard bridge was a tubular structure like the Conway and Britannia bridges also designed by Robert Stephenson. It carried the Grand Trunk Railway of Canada over the St. Lawrence River at Montreal.

1859, ROYAL ALBERT BRIDGE
This was Isambard Kingdom Brunel's last masterwork before his death. The curved, upper chord of oval wrought iron tubes formed a lenticular truss (or Pauli truss as it was known in Europe) of two 445-foot spans over the Tamar in southwest England.

1862, CONCENTRATED ENGINE AND TENDER AXLE LOADS FIRST USED IN COMPUTING BRIDGE LOADINGS
Proposed by Charles Hilton, bridge engineer for the New York Central Railroad, it superseded uniform load per linear foot limits first described by Whipple in 1847. It became standard American engineering practice.

1863, TESTING MACHINE INAUGURATED BY PENNSYLVANIA RAILROAD, ALTOONA SHOPS
A 500-ton-capacity machine designed by William Sellers of Philadelphia tested

Era of the "American Standard" Bridge

By the mid-1870s, economics, materials, and engineering practice had evolved a truss bridge type recognized here and abroad as the "American Standard." There were a variety of truss types, but the Whipple-Murphy was preferred by most engineers. All were fabricated out of pin-connected eye-bars and built-up members made from angles, channels, Ts, I-beams, other standardized rolled sections, and plates riveted together. Wrought iron, almost exclusively, was the material used. The advantages of this system over others was the low cost induced by shop fabrication, shipping parts to the site preassembled, and the speed of erection with semiskilled labor.

This section includes a selection of bridges manufactured by the Berlin Iron Bridge Co., New England's most important structural iron and steel fabricator in the late 19th century. Based in East Berlin, Conn., the company specialized almost exclusively in the lenticular truss, a form first illustrated by Faustus Verantius in the 16th century. George Stephenson's Gaunless Viaduct of 1823, on the Stockton & Darlington Railroad, consisted of three 12-1/2-foot spans, with a fourth being added in 1824 following a flood. Georg Ludwig Frederich Laves of Germany promoted the idea with such persistence that the lenticular truss was known all over eastern Europe as the Laves beam. Isambard Kingdom Brunel built two, his most famous being the Royal Albert Bridge (1859) across the Tamar on the Cornish-Devon border in southwest England.

Lenticular trusses were developed in America by William O. Douglas of Binghamton, N.Y., a graduate of West Point who patented them in 1878 and 1885. The Corrugated Metal Company, a manufacturer of roofing materials and iron trusses, which were necessary to support the heavy corrugated roofing, was the predecessor of Berlin Iron Bridge. The firm, threatened with bankruptcy in 1877, was joined by Douglas who brought in the bridge line. To help manage production, Douglas hired Charles Jarvis, a graduate of Yale University. The firm aggressively marketed and built hundreds of these patented spans from New England to Texas. So successful were the bridges that the name was changed from Corrugated Metal to Berlin Iron Bridge in 1883.

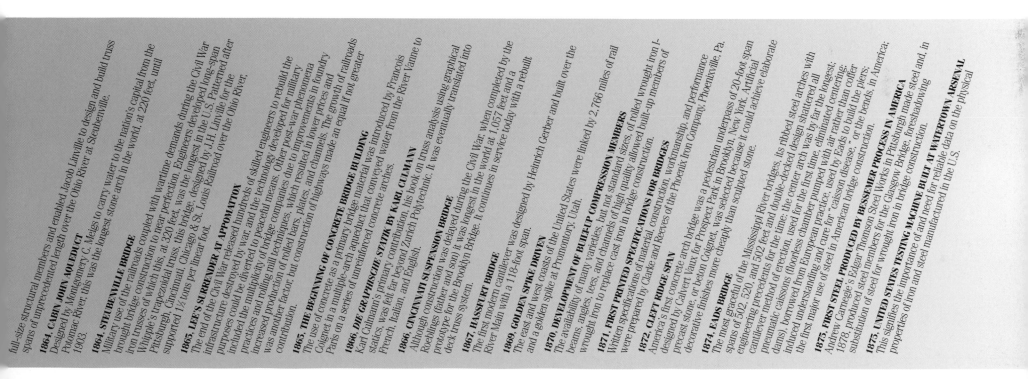

full-size structural members and enabled Jacob Linville to design and build truss spans of unprecedented length over the Ohio River at Steubenville.

1864, CABIN JOHN AQUEDUCT
Designed by Montgomery C. Meigs to carry water to the nation's capital from the Potomac River, this was the longest stone arch in the world, at 220 feet, until 1903.

1864, STEUBENVILLE BRIDGE
Military use of the railroads coupled with wartime demands brought bridge construction to near perfection. Engineers developed long-span iron trusses of which this, at 320 feet, was the longest in the U.S. Patterned after Whipple's trapezoidal truss, this bridge, designed by J.H. Linville over the Ohio River, Pittsburgh, Cincinnati. supported 1½ tons per linear foot.

1865, LEE'S SURRENDER AT APPOMATTOX
The end of the Civil War released hundreds of skilled engineers to rebuild the infrastructure destroyed during the Civil War. Engineers developed long-span trapezoidal truss, at 320 feet, Chicago & St. Louis Railroad over the Ohio River, purposes could be diverted to peaceful means. Other post-war phenomena included the multiplicity of bridge companies due to improvements in foundry practices, which resulted in lower prices and increased production of rolled bars, plates, and channels. The growth of railroads was another factor, but construction of highways made an equal if not greater contribution.

1865, THE BEGINNING OF CONCRETE BRIDGE BUILDING
The use of concrete as a primary bridge material was introduced by François Coignet in a multiple-arch aqueduct that conveyed water from the River Vanne to Paris on a series of unreinforced concrete arches.

1866, DIE GRAPHISCHE STATIK BY KARL CULMANN
Karl Culmann's primary contribution, his book on truss analysis using graphical statics, was felt far beyond Zurich Polytechnic. It was eventually translated into French, Italian, and English.

1866, CINCINNATI SUSPENSION BRIDGE
Although construction was delayed during the Civil War, when completed by the Roeblings (father and son) it was longest in the world at 1,057 feet and a prototype for the Brooklyn Bridge. It continues in service today with a rebuilt deck truss system.

1867, HASSFURT BRIDGE
The first modern cantilever was designed by Heinrich Gerber and built over the River Main with a 118-foot span.

1869, GOLDEN SPIKE DRIVEN
The east and west coasts of the United States were linked by 2,766 miles of rail and a golden spike at Promontory, Utah.

1870, DEVELOPMENT OF BUILT-UP COMPRESSION MEMBERS
The availability of many varieties, but not standard sizes, of rolled wrought iron I-beams, angles, tees, and channels allowed built-up members of wrought iron to replace cast iron in bridge construction.

1871, FIRST PRINTED SPECIFICATIONS FOR BRIDGES
Written specifications of material, construction, workmanship, and performance were prepared by Clarke and Reeves of the Phoenix Iron Company, Phoenixville, Pa.

1872, CLIFFT RIDGE SPAN
America's first concrete arch bridge was a pedestrian underpass for Prospect Park in Brooklyn, New York. Artificial designed by Calvert Vaux for Prospect Park in Brooklyn, New York. Artificial precast stone, or beton Coignet, was selected because it could achieve elaborate decorative finishes more cheaply than sculpted stone.

1874, EADS BRIDGE
The most graceful of the Mississippi River bridges, its ribbed steel arches with spans of 502, 520, and 502 feet and double-decked design shattered all engineering precedents for the time: the center arch was by far the longest; cantilever method of erection, used for the first time, eliminated centering; pneumatic caissons (floorless chamber pumped with air rather than coffer dams), borrowed from European practice, used by Eads to build the piers; induced understanding of "caisson disease," or the bends, in America.

1875, FIRST STEEL PRODUCED BY BESSEMER PROCESS IN AMERICA
Andrew Carnegie's Edgar Thomson Steel Works in Pittsburgh made steel and, in 1878, produced steel members for the Glasgow Bridge, foreshadowing substitution of steel for wrought iron in bridge construction.

1875, UNITED STATES TESTING MACHINE BUILT AT WATERTOWN ARSENAL
This signifies the importance of and need for reliable data on the physical properties of iron and steel manufactured in the U.S.

By the mid-1870s, the era of composite cast and wrought iron bridge construction was over. The 1870s witnessed the gradual elimination of cast iron from bridge construction because wrought iron handled both tensile and compressive stresses and the production costs of wrought iron decreased substantially. The U.S. Army ordered this three span, 400-foot-long bowstring arch truss from the King Iron Bridge & Manufacturing Company of Cleveland. It remains today, in mint condition under National Park Service stewardship, as an engineering feature of this Oregon Trail historic site.

Fort Laramie Bowstring Arch-Truss Bridge

(1875)
North Platt River, Fort Laramie National Historic Site, Wyoming.
Zenas King, Designer.

The Wrought Iron Bridge Company (WIBCo) of Canton, Ohio, was a major competitor of the King Iron Bridge & Manufacturing Company of Cleveland. This bridge features a distinctive upper-chord section, different from those used by King or WIBCo. It is an "Oval Wrought Iron Tubular Arch," patented in 1867 by William Rezner, a physician, John Glass, a foundryman, and George Schneider, a railroad machinist—all from Cleveland. The reason for the variety of upper-chord sections in bowstring truss design was to generate a section that was as strong laterally as vertically to resist the tendency of these members to twist sideways when loaded.

White Bowstring Arch-Truss Bridge

(1877)
Cemetery Drive over Yellow Creek, Poland, Ohio.
William Rezner, John Glass, and
George Schneider, Designers.

The Wrought Iron Bridge Company of Canton, Ohio, also built this remarkable structure, along with hundreds of others throughout the midwest. Remarkable for its span at 302 feet and its 40-foot height for a parallel-chord vehicular truss, it is also the only known example of a triple-intersection Pratt truss (diagonals cross three panel points) in the United States.

Laughery Creek Bridge

(1878)
Old State Route 56 over Laughery Creek,
Aurora vicinity, Indiana.
David A. Hammond, Builder;
Job Abbott, Engineer.

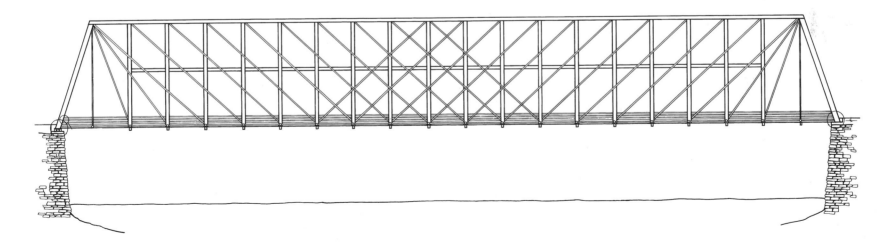

Fourteen bowstring truss bridges survive in Iowa—more than in any other state—because the popularity of this type of truss peaked at a time when the counties in this part of the country were upgrading their road systems. This combination of events and the state's rural character favor the retention of old bridges. The Lower Plymouth Rock Bridge was the oldest bowstring arch in the state until 1986, when it was destroyed.

Lower Plymouth Rock Bridge

(1877, demolished 1986)
Unnumbered Winneshiek County Road
over Upper Iowa River, Kendallville vicinity, Iowa.
David A. Hammond, Builder;
Job Abbott, Engineer.

Freeport Bridge

Freeport Bridge is an example of the Wrought Iron Bridge Company's most popular bridge type as it was the most commonly specified span range, 140–180 feet, in county bridge construction. It uses the "column, plate, and channel" design patented by David Hammond, founder of WIBCo. The Freeport Bridge was relocated to a county park alongside State Route 9 east of Decorah, Iowa, because it is the longest known WIBCo bowstring arch-truss (159 feet 8 inches) in the country.

(1878)
Unnumbered Winneshiek County Road
over Upper Iowa River, Decorah vicinity, Iowa.
David A. Hammond, Builder;
Job Abbott, Engineer.

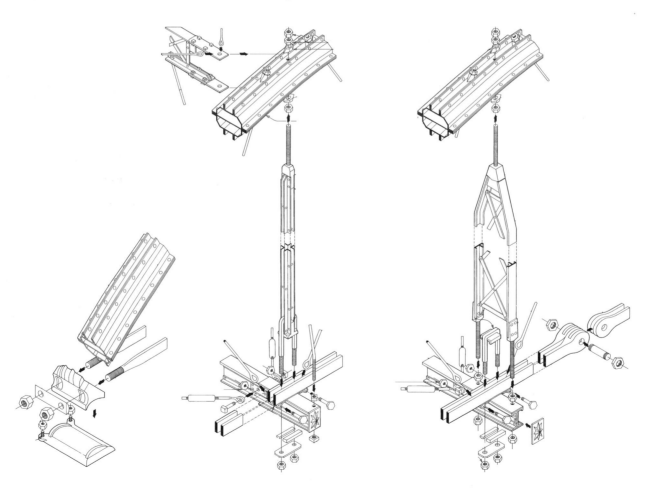

_S_carlets Mill Bridge is a cast and wrought iron Pratt-type truss with an upper chord formed in the shape of an ellipse. It is believed to have been built in the Pottstown Shops of the Reading Railroad under the supervision of John L. Foreman, master carpenter, and used as an overhead crossing on a Reading branch just west of Reading. It was moved to its present site between 1907 and 1935. The bridge is unusual in its use of the elliptical form and the use of cast and wrought iron for its principal structural members at such a late date.

Scarlets Mill Bridge

_(1881)_
Horse-Shoe Trail over Reading Railroad
(abandoned), Scarlets Mill, Pennsylvania.
John L. Foreman, Designer.

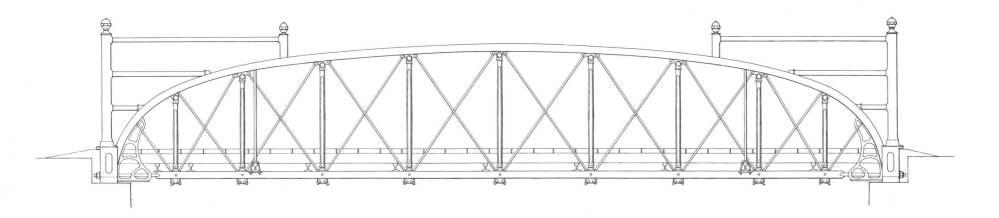

Hayden Bridge

(1882)
*Weyerhaeuser Company rail line
(abandoned) over McKenzie River,
Springfield vicinity, Oregon.
Adolphus Bonzano, Chief Engineer.*

The Hayden Bridge features two milestones of the "American Standard" period of bridge building: ascendancy of the Whipple-Murphy as the truss of choice for intermediate spans and the Phoenix column. Although other truss types were available, the Whipple-Murphy combined the greatest number of advantages, addressing many design circumstances for both rail and vehicular bridges. Clark, Reeves & Company, proprietors of the Phoenixville Bridge Works of Phoenixville, Pennsylvania, was one of the giant bridge manufacturing companies in the east, building hundreds of bridges using the Phoenix column as the primary compression member. The Phoenix section was extremely efficient and the decorative castings in the portals, crested nameplate and finials gave the Phoenix bridge a distinctive touch.

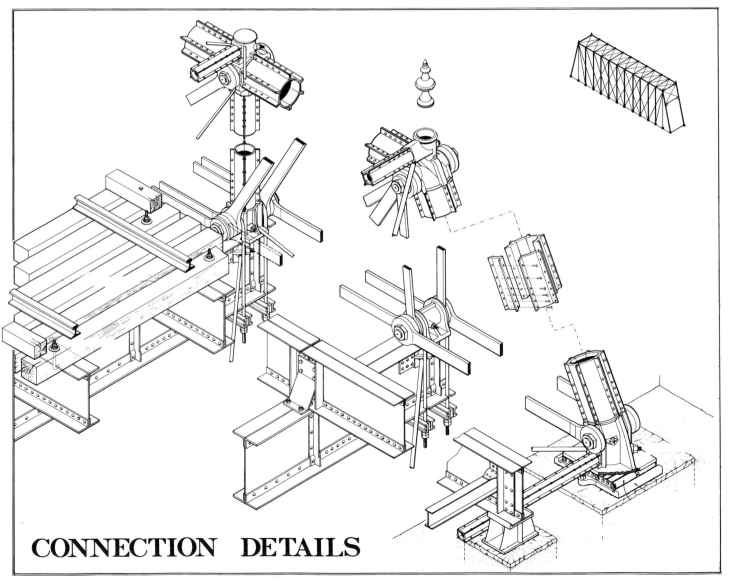

CONNECTION DETAILS

81

This bridge is an example of late 19th century European design and technology much like some of the spans over the Rhine in Germany. Smithfield Street Bridge was the first major commission of Gustav Lindenthal, one of many German engineers who emigrated to the United States, bringing their talents and technology. Using what was known as a Laves beam or Pauli truss in Europe and lenticular here, this bridge is composed of two 360-foot spans. A third line of trusses was added in 1891 when a streetcar line was extended across the river. The distinctive cast steel portals were added in 1915. Scheduled for rehabilitation in 1992–93, the plans call for the preservation or careful replication of the original features and appearance.

Smithfield Street Bridge

(1883, 1891)
Smithfield Street over Monongahela River,
Pittsburgh, Pennsylvania.
Gustav Lindenthal, Engineer.

The truss configuration of this bridge is unique to Ohio. It was manufactured by the Hocking Valley Bridge Company of Lancaster and resembles a patent granted to Archibald McGuffie in 1861. This bridge is a combination of a single-intersection Pratt and a lenticular truss, the deck being suspended from eye-bars. Two other bridges of this configuration survive in Ohio; they are of composite wood and iron construction and covered.

John Bright No. 1 Iron Bridge

(1885)
Havensport Road over Poplar Creek,
Carroll, Ohio.
August Borneman, Designer.

*T*he ornate maker's plaque and cresting that adorned the portals distinguished a Berlin Iron Bridge Company product from that of other bridge manufacturers. Their distinctive appearance and an advantage in costs—the company claimed that the lenticulars used 10 percent less metal than other truss spans—were the principal selling points. The river pier displays the talents of local artisans and supports New York State's only double lenticular, of 341-foot 7-inch span.

Ouaquaga Bridge

(1888)
Dutchtown Road over Susquehanna River,
Ouaquaga, New York.
William O. Douglas, Engineer.

The home state of the Berlin Iron Bridge Company has six surviving lenticular truss bridges. Lenticulars were at the peak of their popularity in the 1880s and 1890s. More than a thousand were built nationwide. Company advertising claimed that the firm had built more than 90 percent of the highway bridges in New England alone.

Lover's Leap Bridge

(1895)
Pumpkin Hill Road
over Housatonic River,
New Milford, Connecticut.
William O. Douglas, Engineer.

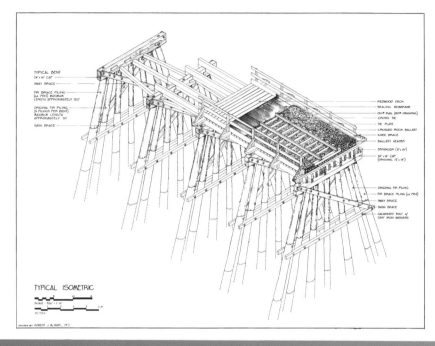

Part Four

TIME LINE 1876–1902

TYPICAL ISOMETRIC

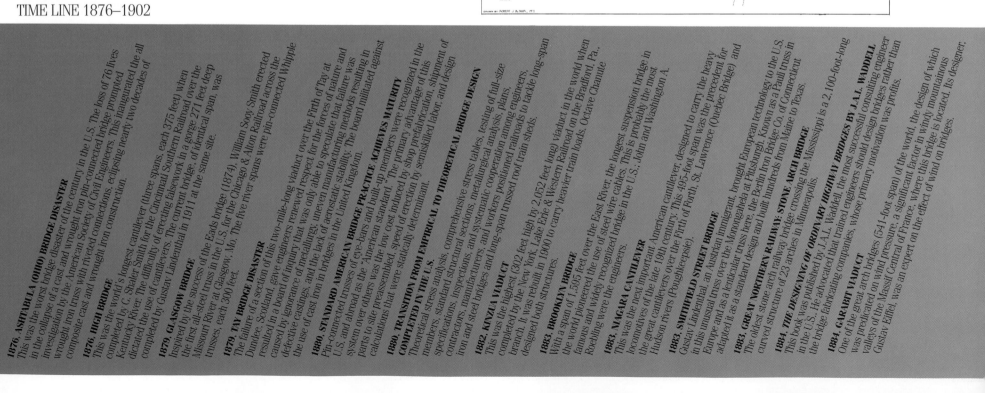

1876, ASHTABULA (OHIO) BRIDGE DISASTER
This was the worst bridge disaster of the century in the U.S. The loss of 76 lives in the collapse of a cast and wrought iron pin-connected bridge prompted an investigation by the American Society of Civil Engineers. This inaugurated the all wrought iron truss with riveted connections, eclipsing nearly two decades of composite cast and wrought iron construction.

1876, HIGH BRIDGE
This was the world's longest cantilever (three spans, each 375 feet) when completed by C. Shaler Smith for the Cincinnati Southern Railroad over the Kentucky River. The difficulty of erecting falsework in a gorge 271 feet deep dictated the use of cantilevers. The current bridge, of identical span, was completed by Gustav Lindenthal in 1911 at the same site.

1879, GLASGOW BRIDGE
Inspired by the success of the Eads bridge (1874), William Sooy Smith erected the first all-steel truss in the U.S. for the Chicago & Alton Railroad across the Missouri River at Glasgow, Mo. The five river spans were pin-connected Whipple trusses, each 300 feet.

1879, TAY BRIDGE DISASTER
The failure of a section of this two-mile-long viaduct over the Firth of Tay at Dundee, Scotland, resulted in a board of inquiry that was only able to speculate that failure was caused by ignorance of metallurgy: uneven manufacturing methods resulting in defective castings, and the lack of aerostatic stability. The board militated against the use of cast iron in bridges in the United Kingdom.

1880, STANDARD AMERICAN BRIDGE PRACTICE ACHIEVES MATURITY
Pin-connected trusses of eye-bars and built-up members were recognized in the U.S. and abroad as the "American standard." The primary advantage of this system over others was the low cost induced by shop prefabrication, shipment of parts to site preassembled, speed of erection by semiskilled labor, and design calculations that were statically determinant.

1880, TRANSITION FROM EMPIRICAL TO THEORETICAL BRIDGE DESIGN COMPLETED IN THE U.S.
Theoretical stress analysis, comprehensive stress tables, testing of full-size members, standard structural sections, metallurgical analysis, plans, specifications, inspections, and systematic cooperation among engineers, contractors, manufacturers, and workers positioned railroads to tackle long-span iron and steel bridges and long-span trussed roof train sheds.

1882, KINZUA VIADUCT
This was the highest (302 feet high) by 2,052 feet long) viaduct in the world when completed by the New York, Lake Erie & Western Railroad on the Bradford, Pa., branch. It was rebuilt in 1900 to carry heavier train loads. Octave Chanute designed both structures.

1883, BROOKLYN BRIDGE
With a span of 1,595 feet over the East River, the longest suspension bridge in the world, this pioneered the use of steel wire cables. This is probably the most famous and widely recognized bridge in the U.S. John and Washington A. Roebling were the engineers.

1883, NIAGARA CANTILEVER
This was the next important American cantilever. This 495-foot span was the precedent for the great cantilevers of the late 19th century.

1883, SMITHFIELD STREET BRIDGE
Gustav Lindenthal, an Austrian immigrant, brought European technology to the U.S. locomotives of the Baltimore & Ohio Railroad designed to carry the heavy in this unusual truss over the Monongahela at Pittsburgh. Known as a Pauli truss in Europe and as a lenticular truss here, the Berlin Iron Bridge Co. of Connecticut adapted it as a standard design and built hundreds from Maine to Texas.

1883, GREAT NORTHERN RAILWAY STONE ARCH BRIDGE
The oldest stone arch railway bridge crossing the Mississippi is a 2,100-foot-long curved structure of 23 arches in Minneapolis.

1884, THE DESIGNING OF ORDINARY HIGHWAY BRIDGES BY J.A.L. WADDELL
This book was published by J.A.L. Waddell, the most successful consulting engineer in the U.S. He advocated that trained engineers should design bridges rather than the bridge fabricating companies, whose primary motivation was profits.

1884, GARABIT VIADUCT
One of the great arch bridges (541-foot span) of the world, the design of which was predicated on wind pressure, a significant factor in windy mountainous valleys of the Massif Central of France, where this bridge is located. Its designer, Gustav Eiffel, was an expert on the effect of wind on bridges.

Great River Crossings

By the 1880s, the transcontinental railroads had been completed. On the older routes, earlier wooden trusses and cast and wrought iron bridges no longer were able to safely carry heavier locomotives and faster trains. New construction combined with massive rebuilding following the Civil War reached a crescendo in the last decades of the 19th century. A similar ongoing process was experienced by cities that had outgrown their infrastructure and needed to upgrade basic services like transportation, water supply, and sewage treatment.

The bridges in this chapter are examples of rail transportation, highway, and civic improvements built during this period. "Great River Crossings" was selected as the category because it captures the optimism that characterized the decades at the turn of the century as the United States completed its manifest destiny of continental settlement. The bridges reflect the material, theoretical, and fabrication improvements that matured during this period.

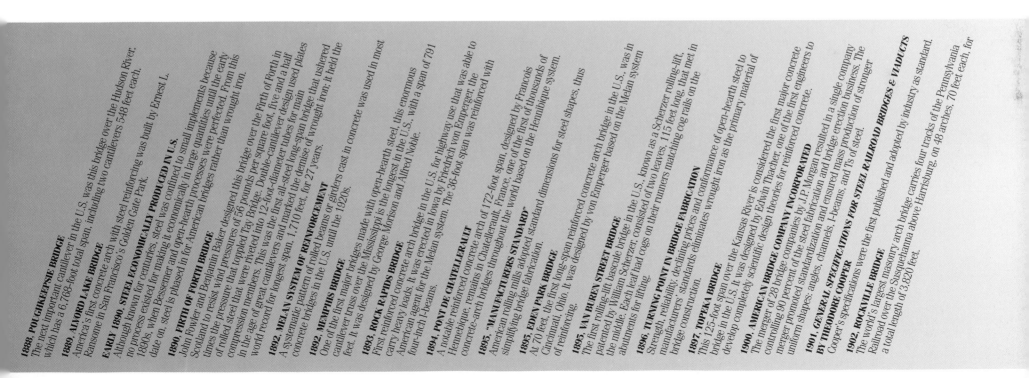

1888, POUGHKEEPSIE BRIDGE
The next important cantilever in the U.S. was this bridge over the Hudson River, which has a 6,768-foot total span, including two cantilevers 548 feet each.

1889, ALVORD LAKE BRIDGE
America's first concrete arch with steel reinforcing was built by Ernest L. Ransome in San Francisco's Golden Gate Park.

EARLY 1890, STEEL ECONOMICALLY PRODUCED IN U.S.
Although known for centuries, steel was confined to small implements because no process existed for making it economically in large quantities until the early 1880s, when Bessemer and open-hearth processes were perfected. From this date on, steel is phased in for American bridges rather than wrought iron.

1890, FIRTH OF FORTH BRIDGE
John Fowler and Benjamin Baker designed this bridge over the Firth of Forth in Scotland to resist wind pressures until the early times of rolled steel for main compression members. Double-cantilever design used plates of rolled steel to resist the pressure that toppled Tay Bridge. Double-cantilever tubes that were riveted into 12-foot-diameter tubes for main compression members. This was the first all-steel long-span bridge that ushered in the age of great cantilevers, 1,710 feet, for 27 years.

1892, MELAN SYSTEM OF REINFORCEMENT
A systematic pattern of rolled beams or girders cast in concrete was used in most concrete bridges in the U.S. until the 1920s.

1892, MEMPHIS BRIDGE
One of the first major bridges made with open-hearth steel, this enormous cantilever truss over the Mississippi is the longest in the U.S. with a span of 791 feet. It was designed by George Morison and Alfred Noble.

1893, ROCK RAPIDS BRIDGE
First reinforced concrete arch bridge in the U.S., for highway use that was able to carry heavy loads. It was erected in Iowa by Friedrich von Emperger, the American agent for the Melan system. The 36-foot span was reinforced with four-inch I-beams.

1894, PONT DE CHATELLERAULT
A notable reinforced concrete arch of 172-foot span, designed by François Hennebique, remains in Chatellerault, France, one of the first of thousands of concrete-arch bridges throughout the world based on the Hennebique system.

1895, "MANUFACTURERS STANDARD"
American rolling mills adopted standard dimensions for steel shapes, thus simplifying bridge fabrication.

1895, EDEN PARK BRIDGE
At 70 feet, the first long-span reinforced concrete arch bridge in the U.S. was in Cincinnati, Ohio. It was designed by von Emperger based on the Melan system of reinforcing.

1895, VAN BUREN STREET BRIDGE
The first rolling-lift bascule bridge in the U.S. was in the Melan system patented by William Scherzer; consisted of two leaves, 115 feet long, that met in the middle. Each leaf had cogs on their runners matching cog rails on the abutments for lifting.

1896, TURNING POINT IN BRIDGE FABRICATION
Strength, reliability, declining prices and conformance of open-hearth steel to manufacturers' standards eliminates wrought iron as the primary material of bridge construction.

1897, TOPEKA BRIDGE
This 125-foot span over the Kansas River is considered the first major concrete bridge in the U.S. It was designed by Edwin Thacher, one of the first engineers to develop completely scientific design theories for reinforced concrete.

1900, AMERICAN BRIDGE COMPANY INCORPORATED
The merger of 28 bridge companies by J.P. Morgan resulted in a single company controlling 90 percent of the steel fabrication and bridge erection business. The merger promoted standardization of uniform shapes: angles, channels, I-beams, and Ts of steel.

1901, GENERAL SPECIFICATIONS FOR STEEL RAILROAD BRIDGES & VIADUCTS BY THEODORE COOPER
Cooper's specifications were the first published and adopted by industry as standard.

1902, ROCKVILLE BRIDGE
The world's largest masonry arch bridge carries four tracks of the Pennsylvania Railroad over the Susquehanna above Harrisburg, on 48 arches, 70 feet each, for a total length of 3,820 feet.

Many historians consider the spanning of the East River the greatest technological feat of the 19th century, and few suspension bridges in the world built since the time of the Roeblings can claim to stand entirely clear of the shadow cast by the Brooklyn Bridge. The plan involved two stone towers, four main cables, anchorages, diagonal stay cables and four trusses that stiffen the deck. The towers stand 276 feet 6 inches above mean high water line and 78 feet 6 inches below on the Manhattan side, 44 feet 6 inches on the Brooklyn side. Because of the great depth below water, they were built on pneumatic caissons—hollow, heavy, timber boxes driven below the river bed. The caissons were forced to the river bottom by successive stone courses laid on their roofs. As the stone was laid on top, men excavated rock, clay, and dirt in pressurized chambers within the caisson. Death and casualties from caisson disease were considerable, even afflicting Washington Roebling, the bridge's builder. It took two years to lay up the 5,434 wires that make up each of the four $15\text{-}^{3}/_{4}$-inch diameter steel cables. Eight 12-foot eye-bars arched 90 degrees provide the transition between cables and anchorages. Each anchorage is a granite and limestone deadweight of 60,000 tons. Diagonal stay cables give the bridge its distinctive appearance, but function to assist the main cables in carrying and stiffening the deck.

Brooklyn Bridge

(1883)
East River,
New York City, New York.
John and Washington A. Roebling, Engineers.

Poughkeepsie Bridge

(1888)
Conrail (abandoned) over Hudson River,
Poughkeepsie, New York.
J.F. O'Rouke, P.P. Dickenson,
and A.B. Paine, Engineers.

Originally built by the Central New England Railroad and leased to the New York, New Haven & Hartford, Poughkeepsie Bridge, with a total length of 6,767 feet, held the record for a steel structure. It has two anchor spans of 525 feet alternating with three cantilevers of 548 feet, two shore spans of 200 feet, and approaches at either end of 3,673 feet. In 1906, the bridge was strengthened by adding a third truss midway between the original two and new columns in the towers. The deck is 212 feet above water and the trusses vary from 37 to 57 feet deep. Conrail closed the bridge in 1974 following a fire and sold it for a dollar to Railway Management Associates. A local citizens group is working to put the bridge back into light-rail and pedestrian service as part of the Hudson Valley Greenway.

Alvord Lake Bridge

(1889)
Haight Avenue entry road over pedestrian walk,
Golden Gate Park,
San Francisco, California.
Ernest L. Ransome, Engineer.

This was America's first concrete arch with steel reinforcing bars. Like other early concrete bridges, it is an ornamental bridge with an imitation-stone finish. Concrete stalactites and stalagmites hang from the arched soffit and sidewalls of this 20-foot span. Ransome was a major proponent of reinforced concrete, one of the most important advances in bridge construction at the turn of the century.

$\mathcal{W}$inona Bridge is a five-span structure consisting of four fixed Parker through-trusses and a single movable Pratt through swing-truss of 440-foot span providing clearance for navigation. The bridge was built by the Union Bridge Company of Athens, Pennsylvania, for the Winona Bridge Railway Company, which was incorporated solely for building and operating this structure.

Winona Bridge

(1891, burned 1989)
Burlington Northern (abandoned) over Mississippi River, Winona, Minnesota.
George S. Morison, Engineer.

At Memphis the Mississippi is a wide, powerful, silt-laden flow that dwarfs all North American rivers. Not since the Eads Bridge in 1874 had anyone succeeded in spanning the lower Mississippi below its confluence with the Missouri. Morison and Noble did it with an asymmetrical cantilever truss of 621-, 621-, 790-, and 226-foot spans, one of the largest of its type when completed. The War Department required the 790-foot navigable-channel span to be just off the Tennessee shoreline. The bridge was so enormous that nine steel companies were required to supply the 8160 tons of steel, which averaged 3-$\frac{1}{2}$ tons per linear foot. The trusses are 30 feet apart, 78 feet deep, and 80 feet above the high water line. So ideal was the crossing at Memphis that two subsequent bridges, the Harahan (1917) and the I-90 (1949), were located adjacent to Morison's bridge.

Memphis Bridge

(1892)

*Kansas City, Fort Scott & Memphis Railroad
over Mississippi River, Memphis, Tennessee.
George S. Morison and Alfred Noble, Engineers.*

This bridge connects Davenport Arsenal, located in the city seen on the other side of the river, and Rock Island Arsenal, seen in the foreground. The bridge, built by the Phoenix Bridge Company, consists of two riveted Pratt through-trusses, five riveted Baltimore through-trusses, and a pin-connected Pratt swing span over the river locks. Altogether, the bridge is nearly 2,000 feet long and carries a roadway, just visible in the photograph, below the double-tracked railroad. The turning drum, rollers, and drive chains for the swing span are still operable today.

Rock Island (Government) Bridge

(1896)

Fort Armstrong Avenue over Mississippi River,
Rock Island Arsenal, Illinois.
Ralph Modjeski, Engineer.

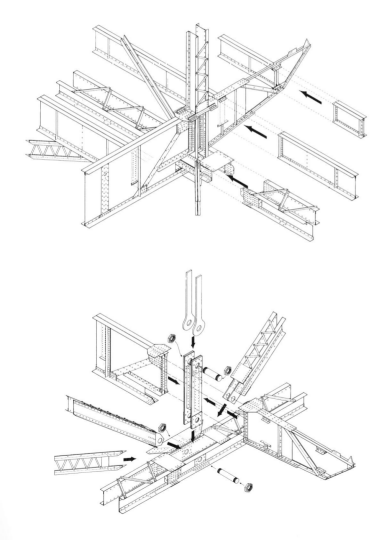

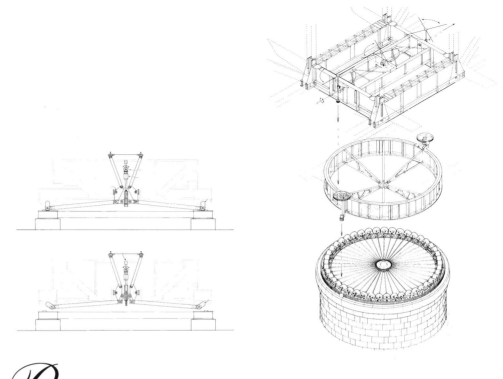

By the 1890s, steel was replacing wrought iron as the material of choice among engineers. Because of its superior strength, it was possible to design bridges of longer spans that were capable of carrying heavier loads. The Pacific Short Line Bridge is one of many erected by the railroads to replace the cast and wrought iron bridges that the first wave of transcontinental railroads built to span the "big rivers:" the Missouri, Mississippi and others that coursed across the great plains of the midwest. Before the Pacific Short Line Bridge was destroyed in 1980, drawings, photographs, and a history were made that documented the character of the "big river" crossings: truss spans in excess of 500 feet, built-up members using standardized steel sections, a combination of riveted and pinned connections, and movable spans providing clearance for navigation.

Pacific Short Line Bridge

(1896, demolished 1980)
US Route 20 over Missouri River,
Sioux City, Iowa.
J.A.L. Waddell, Engineer

The verdant lushness of Kinzua country in northwestern Pennsylvania is maintained today by state parks and national forests. In the late 19th century, however, the landscape yielded minerals and timber in such abundance that the Erie Railroad built this imposing viaduct to haul these materials to market. Viaducts are bridges in which a series of spans are borne on individual towers composed of two or more bents braced together. The first Kinzua Viaduct (1882) was a light iron structure 302 feet high and 2,052 feet long, built by the Phoenixville Bridge Works. It was rebuilt eighteen years later for heavier loads, and continued in service until abandoned in the late 1960s for use as the main feature of Kinzua Bridge State Park. Octave Chanute designed both structures.

Kinzua Viaduct

(1900)
Former Erie Railroad (Bradford Division)
over Kinzua Creek Valley, Kushequa, Pennsylvania.
Adolphus Bonzano and Octave Chanute,
Engineers.

$\mathcal{C}$leveland's historic industrial district is known as "the Flats," an area along the Cuyahoga, a meandering river that served steel mills, factories, and warehouses. A textbook of movable bridges developed along its course, the oldest being this example of a "bobtail" swing. Use of a bobtail allowed the turning drum to rest onshore, not obstructing navigation, while the asymmetrical arrangement of the swing span allowed sufficient length to clear the river. The longer river span was counterbalanced by weights added to the shorter land span. This rare example of a movable bridge was built by the King Bridge Company, one of the prominent bridge fabricators of the late 19th century.

Center Street Bridge

(1901)
Center Street over Cuyahoga River,
Cleveland, Ohio.
James Ritchie and James T. Pardee, Engineers.

The original alignment of the first transcontinental railroad was shortened 43 miles by building the world's longest railway water crossing–a 20-mile-long timber trestle–straight across the Great Salt Lake. Consisting of 38,256 piles, the trestle had a ballast deck so passengers' sleep remained undisturbed when the tempo changed from terra firma to trestle at night. The trestle, raised once, in 1910, needs to be raised again to escape the ever-rising lake level caused by increased runoff during the 1980s. Rebuilt and repaired many times, the original trestle was designed by William Hood.

Ogden–Lucin Cutoff Trestle

(1904)
Southern Pacific Railroad over Great Salt Lake,
Promontory, Utah.
William Hood, Engineer.

Bellows Falls Arch Bridge

(1905, demolished 1982)
Connecticut River, Bellows Falls, Vermont–North Walpole, New Hampshire.
J.R. Worcester, Engineer.

This 540-foot arch was the longest arch span in America when completed. Worcester borrowed from European precedent in selecting an arch for the crossing. Three-hinged trussed arches with suspended decks were not common in the United States. Site conditions and Worcester's ability to demonstrate the cost-effectiveness of this design over conventional bridge types, such as trusses, cantilevers, and suspensions, resulted in this novel engineering solution. The two riverside communities built the bridge to stimulate economic development. Ironically, their economic demise 77 years later contributed to the bridge's demolition in 1982 when the two towns could no longer afford maintenance costs.

$\mathcal{A}$n example of a "City Beautiful" structure, the Taft Memorial Bridge was built during an era following the Chicago Worlds Fair of 1893. The fair influenced architects and engineers to design major civic improvements in a classic Roman style. The bridge has five semicircular arches of 150 feet, and two of 82 feet, with a length between abutments of 1,036 feet and a total length of 1,314 feet. The main arches are hingeless (without reinforcing), but the smaller spandrel arches contain steel. Because appearance was important, the face of the concrete arch rings, pier corners, and trimmings are molded to look like stone.

Taft Memorial Bridge

(1907)
Connecticut Avenue over Rock Creek Park,
Washington, D.C.
George S. Morison, Engineer.

*M*anhattan Bridge is the next crossing of the East River north of the Brooklyn Bridge, linking Manhattan with Brooklyn neighborhoods that burgeoned at the turn of the century. The classically decorative steel towers, designed by Carrere & Hastings, architects, over which the cables are suspended, support dual levels of deck that carry both vehicular and subway traffic. This bridge has a main span of 1,480 feet.

Manhattan Bridge

(1909)
Spanning East River at Pike Street,
New York City, New York.
O. F. Nichols, Engineer;
Carrère & Hastings, Architects.

$\mathcal{Q}$ueensboro is third in a family of five East River bridges linking Manhattan with its boroughs. Gustav Lindenthal, chief engineer of the City of New York's Bridge Department, selected a continuous cantilever for this bridge. Placing piers on Blackwell's Island negated the need for a long suspension span. The Queensboro was the longest cantilever in the United States when completed. It has unequal channel spans of 1,182 feet and 984 feet on either side of the island, anchor spans of 630 feet, and shore arms of 469 feet and 459 feet, for a total length of 3,825 feet 6 inches.

Queensboro Bridge

(1909)
Over East River at 59th Street,
New York City, New York.
Gustav Lindenthal, Engineer.

This is one of the major bridges in Alaska, designed by A.C. O'Neel for the CR&NW Railway, on the route to the gold fields on the Yukon River. It was never repaired after one of the end spans dropped off its pier in 1964 during an earthquake.

Million Dollar Bridge

(1911)
Copper River & Northwest Railway
(abandoned) over Copper River,
Cordova vicinity, Alaska.
A.C. O'Neel, Engineer.

This photograph captures the essence of the Scherzer rolling lift bridge—two riveted circular segmental girders and the counterweights that balance the draw span. The concept was developed by William Scherzer of Chicago, who received 12 patents for variations and improvements on the design between 1893 and 1921. The bridge was fabricated and erected by the Phoenixville Bridge Company, Phoenixville, Pennsylvania. The lifting mechanism is located on the machinery platform seen at the top of the fixed span (at the railing) and consists of two electric motors linked to reduction gears and pinions that engage a horizontal rack that extends the length of the platform. The counterweight, 1,066,000 pounds of concrete plus some loose pig iron, balanced the 187-foot 5-inch Warren truss lift span. The Seddon Island Bridge was destroyed in 1982 when the island was developed.

Seddon Island Bridge

(1909, destroyed 1982)
Seaboard Airline Railway (abandoned) over
Garrison Channel, Tampa, Florida.
William Scherzer, Engineer.

$\mathcal{M}$any bridges built in the early decades of the 20th century were double-decked structures, but the Steel Bridge is unique in having a 211-foot through double-deck vertical lift truss of telescoping design. Waddell & Harrington invented and patented the design for the lift–span. The lower railroad deck, which sees infrequent service, can be raised independently of the upper vehicular deck, which usually has a constant stream of traffic. However, both decks can be raised together, allowing a tall ship to pass.

Steel Bridge

(1912)
Union Pacific & Southern Pacific RRs
over Willamette River, Portland, Oregon.
Waddell & Harrington, Engineers.

he abruptly rising hills in the background capture the high plains of the western United States where this bridge is located. The bridge was among the longest vertical lift bridges in the world, composed of three riveted Parker through trusses, 275 feet each, and a 296-foot vertical lift span. It was modified for highway use in 1926 when the Missouri began to collect silt and steamboat traffic waned.

Snowden Lift Bridge

(1913)
Burlington Northern Railway (abandoned)
over Missouri River,
Snowden, Montana.
Waddell & Hardesty, Engineers.

Since this bridge's demolition, its cast iron ornaments have been displayed at the headquarters of the Pittsburgh History & Landmarks Foundation. The superstructure consisted of two subdivided Baltimore trusses, 531 feet long and 36 feet wide. The original design called for stone, but portals of cast iron, steel, and bronze were substituted for reasons of cost. Various designs were developed by Stanley L. Roush, architect, in collaboration with sculptor Charles Keck, of New York. Shown kneeling on either side of the Arms of the City are Christopher Gist, civil engineer, surveyor, and pioneer, and Guyasuta, a local Indian chief. The north (Manchester) portal had a coal miner and steelworker on either side of the municipal escutcheon.

Manchester Bridge

(1915, demolished 1970)
Spanning Monongahela River,
Pittsburgh, Pennsylvania.
Department of Public Works, City of Pittsburgh,
Joseph W. Armstrong, Director, Designer;
Emil Swensson, Consulting Engineer;
Stanley L. Roush, Architect.

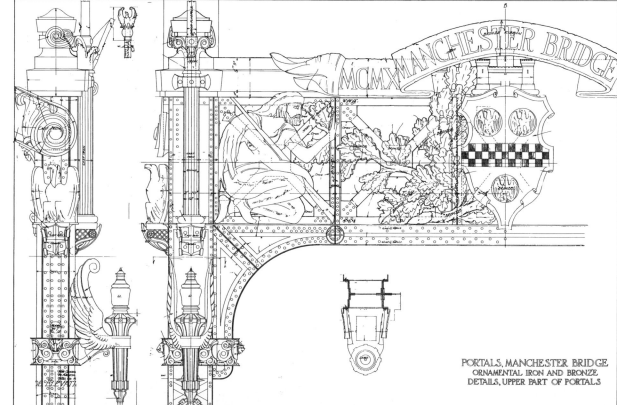

PORTALS, MANCHESTER BRIDGE
ORNAMENTAL IRON AND BRONZE
DETAILS, UPPER PART OF PORTALS

Nicholson is dwarfed by this 240-foot-high viaduct, a massive reinforced concrete arch structure—the largest in the world. It was designed by A. Burton Cohen as part of a system-wide improvement to upgrade the DL&W with permanent low-maintenance concrete structures and to create a new alignment and grade, the Clarks Summit–Hallstead Cutoff, that shortened travel time. The 2,375-foot-long structure is composed of ten semicircular, two-ribbed, open-spandrel arches of 180-foot span each.

Tunkhannock Viaduct

(1915)
*Delaware, Lackawanna & Western Railroad
over Tunkhannock Creek,
Nicholson, Pennsylvania.
George J. Ray, Chief Engineer;
A. Burton Cohen, Design Engineer.*

Hell Gate Bridge

(1917)

New York Connecting Railroad over
East River at Hell Gate,
Queens and The Bronx, New York.
Gustav Lindenthal, Engineer.

Awesome in appearance, weight, and span, the Hell Gate Bridge was designed by one of the master bridge builders of the 20th century, Gustav Lindenthal. He selected a two–hinge spandrel-braced arch to link the New York, New Haven & Hartford Railroad and its connecting lines with the Pennsylvania Railroad and its connecting lines. Spanning 977 feet from center to center of pins, 80,000 tons of steel were required to support a combined dead and live load of 75,000 pounds per linear foot. Only the lower chord of the arch carries these loads, while the upper chord serves as the top member of a stiffening truss. Massive stone abutments and portal towers complete the monumentality of this impressive bridge. The approach spans on both sides of the water are extensive.

Northern New Jersey Bridges

Northern New Jersey has an impressive number of bridges. Four railroads, one rapid-transit system, and numerous highways require more than two dozen bridges, most of them movable spans, over 11 waterways. More than half of them span the Hackensack and Passaic rivers. Grouped with the tank farms, generating plants, and truck depots, they create an industrial landscape that cannot be missed from the I-95 viaducts over New Jersey's meadowlands.

(Over Hackensack River in order of recession), Kearny-Jersey City, New Jersey.

1. *New Jersey Transit (formerly Erie-Lackawanna Railroad): Lower Hackensack Lift Bridge.*
2. *Newark Turnpike Lift Bridge.*
3. *Conrail (formerly Penn Central Railroad): Hack Lift Bridge.*
4. *Port Authority (PATH) Lift Bridge.*
5. *Pulaski Skyway.*
Various engineers, dates, and builders.

121

The uniqueness of this bridge derived from extreme tidal conditions that precluded ordinary solutions. Surveys revealed that a ledge just a few feet underwater extended nearly the entire length between Orrs and Bailey islands, rather than a tidal channel. Llewellyn Nathaniel Edwards, bridge engineer and author of the classic bridge study, *A Record of History & Evolution of Early American Bridges*, decided to construct the bridge on the rock ledge, hence the sinuous alignment seen in the photograph. The bridge is laid up, like an open crib dam, of granite slabs. The cellular construction permits the tide to ebb and flow freely while at the same time is heavy enough to withstand the hammering waves and winds of a New England northeaster. The crib structure has a total length of 1,120 feet, including a fixed channel span of 52 feet and a width of 18 feet curb to curb.

Bailey Island Bridge

(1928)
State Route 24 over Casco Bay,
Harpswell, Maine.
Llewellyn Nathaniel Edwards, Engineer.

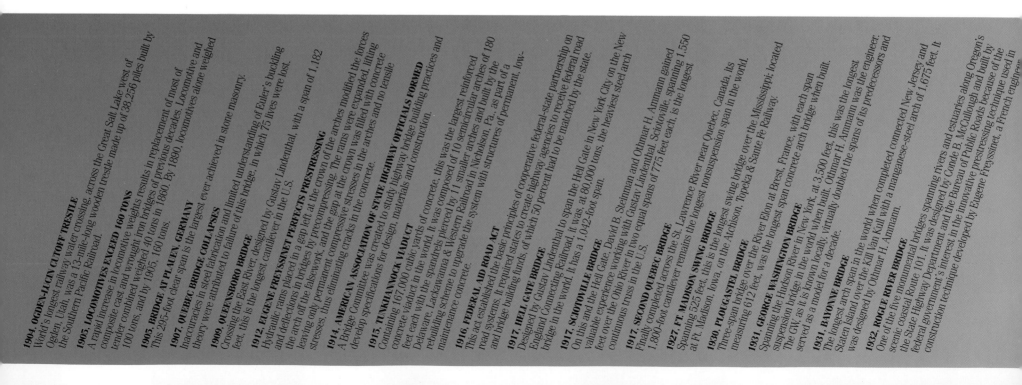

Part Five

TIME LINE 1904–1991

1904, OGDEN-LUCIN CUTOFF TRESTLE
World's longest railway water crossing, across the Great Salt Lake west of Ogden, Utah, was a 13-mile-long wooden trestle made up of 38,256 piles built by the Southern Pacific Railroad.

1905, LOCOMOTIVES EXCEED 160 TONS
A rapid increase in locomotive weights results in replacement of most of composite cast and wrought iron bridges of previous decades. Locomotive and tender combined weighed 40 tons in 1860. By 1890, locomotives alone weighed 100 tons, and by 1905, 160 tons.

1905, BRIDGE AT PLAUEN, GERMANY
This 295-foot clear span is the largest ever achieved in stone masonry.

1907, QUEBEC BRIDGE COLLAPSES
Inaccuracies in steel fabrication and limited understanding of Euler's buckling theory were attributed to failure of this bridge, in which 75 lives were lost.

1909, QUEENSBORO BRIDGE
Crossing the East River, designed by Gustav Lindenthal, with a span of 1,182 feet, this is the longest cantilever in the U.S.

1912, EUGENE FREYSSINET PERFECTS PRESTRESSING
Hydraulic rams placed in a gap left at the crown of the arches modified the forces and deflections in bridges by precompressing. The rams were expanded, lifting the arch off the falsework, and the gap at the crown was filled with concrete leaving only permanent compressive stresses in the arches and no tensile stresses, thus eliminating cracks in the concrete.

1914, AMERICAN ASSOCIATION OF STATE HIGHWAY OFFICIALS FORMED
A Bridge Committee was created to study highway bridge building practices and develop specifications for design, materials and construction.

1915, TUNKHANNOCK VIADUCT
Containing 167,000 cubic yards of concrete, this was the largest reinforced concrete viaduct in the world. It was composed of 10 semicircular arches of 180 feet each with the spandrels pierced by 11 smaller arches and built by the Delaware, Lackawanna & Western Railroad in Nicholson, Pa., as part of a rebuilding scheme to upgrade the system with structures of permanent, low-maintenance concrete.

1916, FEDERAL AID ROAD ACT
This act established the basic principles of cooperative federal-state partnership on road systems. It required states to create highway agencies to receive federal road building funds, of which 50 percent had to be matched by the state.

1917, HELL GATE BRIDGE
Designed by Gustav Lindenthal to span the Hell Gate in New York City on the New England Connecting Railroad, it was, at 80,000 tons, the heaviest steel arch bridge in the world. It has a 1,042-foot span.

1917, SCIOTOVILLE BRIDGE
On this and the Hell Gate, David B. Steinman and Othmar H. Ammann gained valuable experience working with Gustav Lindenthal. Sciotoville, spanning 1,550 feet over the Ohio River in two equal spans of 775 feet each, is the longest continuous truss in the U.S.

1917, SECOND QUEBEC BRIDGE
Finally completed across the St. Lawrence River near Quebec, Canada, its 1,800-foot cantilever remains the longest nonsuspension span in the world.

1927, FT. MADISON SWING BRIDGE
Spanning 525 feet, this is the longest swing bridge over the Mississippi; located at Ft. Madison, Iowa, on the Atchison, Topeka & Sante Fe Railway.

1930, PLOUGASTEL BRIDGE
Three-span bridge over the River Elon at Brest, France, with each span measuring 612 feet, was the longest span concrete arch bridge when built.

1931, GEORGE WASHINGTON BRIDGE
Spanning the Hudson River in New York, at 3,500 feet, this was the longest suspension bridge in the world when built. Othmar H. Ammann was the engineer. The GW, as it is known locally, virtually doubled the spans of its predecessors and served as a model for a decade.

1931, BAYONNE BRIDGE
The longest arch span in the world when completed connected New Jersey and Staten Island over the Kill Van Kull with a manganese-steel arch of 1,675 feet. It was designed by Othmar H. Ammann.

1932, ROGUE RIVER BRIDGE
One of the five monumental bridges spanning rivers and estuaries along Oregon's scenic coastal Route 101. It was designed by Conde B. McCullough and built by the State Highway Department and the Bureau of Public Roads because of the federal government's interest in the innovative prestressing technique used in construction technique developed by Eugene Freyssinet, a French engineer.

Modern Developments

The following Oregon bridges are grouped because they represent a remarkable outpouring of creativity and skill matched by few state highway departments in the country. Conde Balcom McCullough was state bridge engineer during the time of the Oregon Coast Highway Commission, which used funds from Franklin Roosevelt's Public Works Administration to complete the Oregon Coast Highway, a scenic route along Oregon's coast. The project provided jobs for the unemployed during the Great Depression. McCullough and his staff of engineers designed and supervised the construction of 10 major bridges in a remarkable five year period.

While each span is different, a common theme of reinforced concrete and steel arches embellished by Art Deco motifs and clean, streamlined lines harmonized this family of bridges with the Oregon coast, one of the most scenic in the country. This family of bridges represents some of the best and most innovative concrete and steel bridge designs in the world.

Engineers like Othmar Ammann in New York and Joseph Strauss in San Franscisco were designing comprehensive road and bridge projects in other parts of the country. The bridges included in this section portray the last epic of monumental bridge building. Few bridges of the scale and costs of the George Washington and the Verrazano will be built in the United States once the interstates are completed by the end of the 20th century.

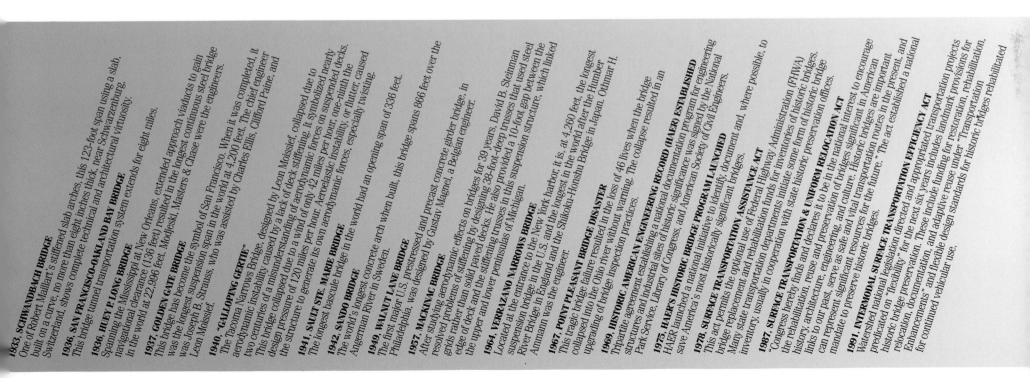

1933, SCHWANDBACH BRIDGE
One of Robert Maillart's stiffened slab arches, this 123-foot span using a slab, built on a curve, no more than eight inches thick, near Schwarzenburg, Switzerland, shows complete technical and architectural virtuosity.

1936, SAN FRANCISCO-OAKLAND BAY BRIDGE
This bridge tunnel transportation system extends for eight miles.

1936, HUEY P. LONG BRIDGE
Spanning the Mississippi at New Orleans, extended approach viaducts to gain navigational clearance (136 feet) resulted in the longest continuous steel bridge in the world at 22,996 feet.

1937, GOLDEN GATE BRIDGE
This bridge has become the symbol of San Francisco. When it was completed, it was the longest suspension span in the world at 4,200 feet. The chief engineer was Joseph B. Strauss, who was assisted by Charles Ellis, Clifford Paine, and Leon Moisseif.

1940, "GALLOPING GERTIE"
The Tacoma Narrows Bridge, designed by Leon Moisseif, collapsed due to aerodynamic instability caused by a lack of deck stiffening. It symbolized nearly two centuries of a misunderstanding of aerodynamic forces on suspended decks. This bridge collapsed due to a wind of only 42 miles per hour, one-ninth the design pressure of 120 miles per hour. Aeroelastic instability, or flutter, caused the structure to generate its own aerodynamic forces, especially twisting.

1941, SAULT STE. MARIE BRIDGE
The longest bascule bridge in the world had an opening span of 336 feet.

1942, SANDO BRIDGE
The world's longest concrete arch when built, this bridge spans 866 feet over the Angerman River in Sweden.

1949, WALNUT LANE BRIDGE
The first major U.S. prestressed and precast concrete girder bridge, in Philadelphia, was designed by Gustav Magnel, a Belgian engineer.

1957, MACKINAC BRIDGE
After studying aerodynamic effects on bridges for 39 years, David B. Steinman resolved problems of stiffening by designing 38-foot-deep trusses that used steel grids rather than solid paved decks. He also provided a 10-foot gap between the edge of deck and the stiffening trusses in this suspension structure, which linked the upper and lower peninsulas of Michigan.

1964, VERRAZANO NARROWS BRIDGE
Located at the entrance to the New York harbor, it is, at 4,260 feet, the longest suspension bridge in the U.S. and the longest in the world after the Humber River Bridge in England and the Shikoku-Honshu Bridge in Japan. Othmar H. Ammann was the engineer.

1967, POINT PLEASANT BRIDGE DISASTER
This tragic bridge failure resulted in the loss of 46 lives when the bridge collapsed into the Ohio river without warning. The collapse resulted in an upgrading of bridge inspection practices.

1969, HISTORIC AMERICAN ENGINEERING RECORD (HAER) ESTABLISHED
Tripartite agreement establishing a national documentation program for engineering structures and industrial sites of historic significance was signed by the National Park Service, Library of Congress, and American Society of Civil Engineers.

1975, HAER's HISTORIC BRIDGE PROGRAM LAUNCHED
HAER launched a national initiative to identify, document, and, where possible, to save America's most historically significant bridges.

1978, SURFACE TRANSPORTATION ASSISTANCE ACT
This act permits the optional use of Federal Highway Administration (FHWA) bridge replacement and rehabilitation funds for inventories of historic bridges. Many state transportation departments initiate some form of historic preservation inventory, usually in cooperation with state historic preservation offices.

1987, SURFACE TRANSPORTATION & UNIFORM RELOCATION ACT
"Congress hereby finds and declares it to be in the national interest to encourage the rehabilitation, reuse and preservation of historic bridges. Historic bridges are important links to our past, serve as safe and vital transportation routes in the present, and can represent our past, serve as safe and vital transportation resources for the future." The act established a national mandate to preserve historic bridges.

1991, INTERMODAL SURFACE TRANSPORTATION EFFICIENCY ACT
Watershed national legislation directed and appropriated funding for significant interest in American history, architecture, engineering, and culture. Historic bridges significant in American predicated on "flexibility" for the next six years includes landmark provisions for historic bridge preservation. These include funding for restoration, rehabilitation, relocation, documentation, and adaptive reuse under "Transportation Enhancements" and flexible design standards for historic bridges rehabilitated for continued vehicular use.

St. Johns Bridge is not one of the Oregon coast bridges, but it is definitely related aesthetically. This is evident in the Gothic arch motifs and other embellishments of this suspension span. Multnomah County contracted with the firm of Steinman and Robinson to design the 1,207-foot suspension span. The 400-foot high Gothic towers carry U.S. Route 30 200 feet above the Willamette River.

St. Johns Bridge

(1931)
US Route 30 over Willamette River,
Portland, Oregon.
David B. Steinman, Engineer.

$\mathcal{R}$ogue River Bridge is one of the larger spans on Oregon's coastal road, with a total length of 1,898 feet consisting of two approach spans and seven 230-foot, two-ribbed, open-spandrel, reinforced-concrete deck arches. McCullough's decorative palette consisted of classical forms—columns, arches and concrete scoured to look like stone, combined with Art Deco and streamlined motifs. McCullough was interested in innovative structural techniques as much as aesthetics. This bridge was the first reinforced-concrete arch built in the United States using the Freyssinet method of construction. The experiment was cosponsored by the Bureau of Public Roads, which hoped to determine the advantages and disadvantages of the Freyssinet system. Essentially the goal was to reduce construction costs as much as 10 percent, by decreasing the amount of reinforcing steel and concrete required. This was accomplished by precompressing the arches, literally jacking them apart before the concrete had a chance to set fully, and then filling in the skewbacks and keys, the hinge points around which the arch could rotate (move), with concrete plugs. This left the arch ribs permanently stressed, minimizing structural and other cracking while the concrete cured. The visual result was thinner elegant arches, and a modest cost savings.

Rogue River Bridge

(1932)
Oregon Coast Highway (US Route 101) over Rogue River, Gold Beach, Oregon.
Conde B. McCullough, Engineer.

127

Cape Creek Bridge

(1932)
Oregon Coast Highway (US Route 101) over
Cape Creek, Florence vicinity, Oregon.
Conde B. McCullough, Engineer.

Cape Creek Bridge is unusual for its double-tiered composition, much like the Pont du Gard, the famous Roman aqueduct in France. Cape Creek forms a narrow precipitous gorge with high banks. Rather than a massive steel arch spanning the valley in a single reach or a series of multiple arches, McCullough chose a modest 220-foot river span and the two-tiered multiple arched viaducts for his approaches, which were in scale with this scenic valley.

Alsea Bay Bridge

(1936, replaced 1992)
Oregon Coast Highway (US Route 101) over
Alsea Bay, Waldport, Oregon.
Conde B. McCullough, Engineer.

Of the 10 major bridges and dozens of smaller ones on the Oregon Coast Highway, Alsea Bay was considered the crown jewel. The bridge, a combination of deck and tied arches, is over 3,000 feet long. Several years ago, deterioration of the concrete in the deck and the spandrel piers alarmed bridge inspectors. Unfortunately, no reasonable means for arresting concrete deterioration is known, and it was replaced in 1992. This photograph shows the piers of the new bridge under construction.

The narrow channel of the Siuslaw made the bascule draw the most desirable choice. The draw span, a 140-foot double-leaf bascule, is flanked by two tied concrete arches, 154 feet each. The massive piers at either end of the draw contain the counterweights. The four houses on top at deck level, with the ornate roofs and sunburst designs cast in the concrete, contain the control mechanisms. Approaches to the arched spans total 20 concrete deck girders, eight on the north side and 12 on the south.

Siuslaw River Bridge

(1936)
Oregon Coast Highway (US Route 101) over
Siuslaw River, Florence, Oregon.
Conde B. McCullough, Engineer.

Umpqua River Bridge

(1936)
Oregon Coast Highway (US Route 101) over
Umpqua River, Reedsport, Oregon.
Conde B. McCullough, Engineer.

This movable bridge consists of a 430-foot steel Parker truss swing span, flanked by two reinforced-concrete through-tied arches each 154 feet in length. The bridge has all the earmarks of a McCullough masterpiece—strong classical forms combined with Art Deco and Modern motifs, concrete pylons at the portal, and an elaborately bracketed balustrade. This underdeck view shows another McCullough signature, the Gothic arch.

Yaquina Bay Bridge

(1936)
Oregon Coast Highway
(US Route 101) over Yaquina Bay,
Newport, Oregon.
Conde B. McCullough, Engineer.

Yaquina Bay Bridge is a combination of steel and reinforced concrete arches with a total length of 3,260 feet. The main channel span is a through arch 600 feet long and 226 feet high, with the road suspended from hangers. It is flanked by two steel deck arches, each 350 feet long. There are 10 ribbed open-spandrel deck arches on either side of the main steel arch span, ranging from 160 to 265 feet long. Approaches to the bridge are concrete girder deck spans.

*C*oos Bay Bridge was one of the last links in the Oregon coast route. The distinctive 739-foot cantilever truss was the structural type McCullough chose to resolve the navigational requirements of Coos Bay harbor, which served oceangoing ships. The balance of the 5,305-foot structure is composed of 13 double-ribbed, open-spandrel deck approach arches. The bridge was named McCullough Memorial Bridge in recognition of the engineer's contribution to Oregon's highway system.

Coos Bay Bridge

(1936)
Oregon Coast Highway (US Route 101)
over Coos Bay
North Bend, Oregon.
Conde B. McCullough, Engineer.

*N*avigation requirements and nature dictated this 1,675-foot parabolic arch span, the longest arch bridge in the world until recently, when the New River Gorge Bridge opened in West Virginia. The waterway, linking the busiest port facilities on the eastern seaboard—Newark Bay and Upper Bay of New York harbor—made a single-span, high-level bridge a necessity. Dense, fine-grained bedrock on both shores assured excellent foundation conditions so Othmar Ammann, bridge engineer for the New York Port Authority, selected an arch rather than a suspension bridge whose anchorages would have been expensive to excavate. New high-strength alloy steel, employing manganese, was used for the first time in the arch ribs and rivets.

Bayonne Bridge

(1931)
State Route 440 over Kill van Kull,
Bayonne, New Jersey,
Port Richmond, Staten Island, New York.
Othmar Ammann, Engineer.

George Washington Bridge

(1931)
Interstate 95 over Hudson River,
New York, New York.
Othmar Ammann, Leon S. Moissief, and
Allston Dana, Engineers;
Cass Gilbert, Architect.

The only type of bridge eminently suited for very long spans of more than 2,000 ft is the suspension bridge. The George Washington Bridge dramatically doubled the spans of its predecessors. The team assembled by the New York Port Authority for this momentous task consisted of engineers Othmar Ammann, Leon Moissief, and Allston Dana and architect Cass Gilbert. The extraordinary magnitude of the bridge is seen in the towers, a dense space frame of trussed bents and arches that carry the eight-lane road and lower deck that was added later. Inside the anchorages, the cables are splayed into individual wire strands. Improvements in steelmaking enabled the unprecedented length of 3,500 feet. Compared with the Brooklyn Bridge, where steel wire was used for the first time at 160,000 pounds per square inch, wire cable strength for the George Washington reached 240,000 pounds per square inch by 1931.

*T*his 6.2-mile viaduct snakes its way from Newark across the Passaic and Hackensack rivers to feed automobiles under the Hudson River via the Holland Tunnel. Shown in this picture are the two steel cantilever through trusses, each 1,250 feet long. One crosses the Passaic, and the other in the distance crosses the Hackensack. The Skyway was one of the first elevated expressway systems.

Pulaski Skyway

(1932)
US Routes 1 & 9 over Passaic and Hackensack
Rivers, Newark & Jersey City, New Jersey.
Sigvald Johannesson, Engineer.

San Francisco— Oakland Bay Bridge

(1936)
Interstate 80 over San Francisco Bay,
Oakland, California.
C.H. Purcell, Glenn B. Woodruff, Ralph
Modjeski, and Leon S. Moissief, Engineers.

Eight miles is the overall length of this bridge-tunnel transportation system built by the California Toll Bridge Authority and the California Department of Public Works at the same time as the Golden Gate Bridge. The bridge consists of two main components. The West Bay crossing, shown in the photograph, connects San Francisco with Yerba Buena Island in the middle of the bay and consists of two suspension spans with a total length of 9,200 feet. It is the only double-span suspension bridge in the world. The East Bay crossing consists of a 1,400-foot cantilever, five through-trusses and 14 deck trusses for a total length of 11,000 feet not counting the Oakland approach spans. Yerba Buena Island is pierced by a 1,800-foot-long tunnel.

Along with the Brooklyn Bridge, this is one of the world's best-known bridges and a civil engineering masterpiece. Under the leadership of Joseph B. Strauss, chief engineer, with the assistance of Charles Ellis and Leon Moissief, the 4,260-foot span superseded the span of the George Washington Bridge and held sway till the Verrazano Narrows Bridge was completed 27 years later. More than 100,000 tons of steel, 693,000 cubic yards of concrete, and 80,000 miles of wire cable were used in the bridge's construction. The major engineering challenge was the foundations for the 746-foot-tall towers. Strong tides aborted several attempts to sink pneumatic caissons into the bay floor, causing the engineers to use cofferdams instead. The Art Deco towers and the iron-oxide red color were the contributions of Irving Morrow, an architect working under Strauss' direction.

Golden Gate Bridge

(1937)
US Route 101 over Golden Gate,
San Francisco, California.
Charles B. Strauss, Charles Ellis, and Leon S. Moissief, Engineers;
Irving Morrow, Architect.

Verrazano Narrows Bridge

(1964)
Interstate 278 over Verrazano Narrows,
New York City.
Ammann & Whitney, Engineers.

The Verrazano Narrows Bridge is the longest span in America and the third longest suspension bridge in the world after the River Humber Bridge in England and the Shikoku-Honshu Bridge in Japan. Designed by Ammann & Whitney, it is 4,260 feet from tower to tower, has a total length of 7,200 feet, and a vertical clearance of 216 feet for ships entering and leaving New York harbor. The massiveness of the structure is apparent from miles away as one views the towers rising above the roofs of Brooklyn and Staten Island. By the time the cables reach the anchorage gallery, the compacted cables are splayed into individual strands that wrap around eye–bars embedded in massive blocks of concrete.

Appendix

The following is a list of the bridges, trestles, aqueducts, viaducts, and other bridge-like structures in the HAER and HABS collections as of March 1992. They are arranged alphabetically by state, city, and structure.

An * indicates that the structure is a member of the HABS collection. All other structures are part of the HAER collection. A + indicates a National Historic Civil Engineering Landmark Bridge, and a # indicates a National Historic Landmark.

Additional information and documentation on these structures can be obtained by visiting or contacting the Library of Congress Prints and Photographs Division, Washington, DC 20540.

ALABAMA
Big Bear Creek Covered Bridge *, Allsboro Vicinity
Cripple Deer Creek Covered Bridge *, Allsboro Vicinity
Bridgeport Swing Span Bridge, Bridgeport Vicinity
Buzzard Roost Covered Bridge *, Cherokee Vicinity
Covered Bridge *, Eastaboga Vicinity
Bridges of the Upper Tombigbee River Valley, Tombigbee Valley Vicinity

ALASKA
Copper River & Northwestern RR: Gilahina Bridge, Chitina Vicinity
Kuskalana Bridge, Chitina Vicinity
Copper River & Northwest RR: Million Dollar Bridge, Cordova Vicinity
Nizina Bridge, McCarthy Vicinity

ARIZONA
Verde River Sheep Bridge, Cave Creek Vicinity
Kaibab Trail Suspension Bridge, Grand Canyon National Park
Holbrook Bridge, Holbrook
Rio Puerco Bridge, Holbrook Vicinity
Arizona Eastern RR Bridge, Tempe
Ash Ave. Bridge, Tempe

ARKANSAS
Augusta Bridge, Augusta
Beaver Bridge, Beaver
Spring Lake Bridge, Belleville Vicinity
Old River Bridge, Benton
Saline River Bridge, Benton
Ouachita River Bridge, Calion
Mountain Fork Bridge, Camp Pioneer Vicinity
Clarendon Bridge, Clarendon
Cotter Bridge +, Cotter
Lee Creek Bridge, Cove City
White River Bridge, De Valls Bluff
Mulladay Hollow Bridge, Eureka Springs Vicinity
Wyman Bridge, Fayetteville
St. Francis River Bridge, Forrest City
South Fork Bridge, Fountain Lake Vicinity
Big Piney Creek Bridge, Ft. Douglas
Red River Bridge, Garland City
Spavinaw Creek Bridge, Gravette Vicinity
Harp Creek Bridge, Harrison Vicinity
Osage Creek Bridge, Healing Springs Vicinity
Winkley Bridge, Heber Springs
Ricks Estate Stone Bridge *, Hot Springs Vicinity
St. Louis–San Francisco Bridge, Imboden
Jenny Lind Bridge, Jenny Lind Vicinity
Judsonia Bridge, Judsonia
St. Francis River Bridge, Lake City
Lincoln Ave. Viaduct, Little Rock
Second Street Bridge, Little Rock

Little Cossatot River Bridge, Lockesburg Vicinity
Rockport Bridge, Malvern
Milltown Bridge, Milltown
Lee Creek Bridge (No. 1), Natural Dam
Newport Bridge, Newport
North Fork Bridge, Norfolk
Edgemere Street Bridge, North Little Rock
Fourteenth Street Bridge, North Little Rock
Lake No. 1 Bridge, North Little Rock
Lakeshore Drive Bridge, North Little Rock
Achmun Creek Bridge, Ola Vicinity
Little Missouri River Bridge, Old Rome Vicinity
Eight Mile Creek Bridge, Paragould
Cypress Creek Bridge, Perry Vicinity
Cedar Creek Bridge, Petit Jean State Park
Black River Bridge, Pocahontas
Buffalo River Bridge, Pruitt
Illinois River Bridge, Siloan Springs
Springfield-Des Arc Bridge, Springfield
Cache River Bridge, Walnut Ridge
War Eagle Bridge, War Eagle

CALIFORNIA
Van Duzen Bridge, Carolotta Vicinity
Crow Creek Bridge, Castro Valley
Salinas River Bridge, Chualar Vicinity
Cache Creek Bridge, Clear Lake
Southern Pacific RR,Colfax Viaduct, Colfax Vicinity
Colusa Bridge, Colusa
Smith River Bridge, Crescent City Vicinity
Dardanelle Bridge, Dardanelle Vicinity
Moody Bridge, Garberville Vicinity
Llagas Creek Bridge, Gilroy Vicinity
Covered Bridge *, Glen Canyon
Bridgeport Covered Bridge * +, Grass Valley
Gianella Bridge, Hamilton City Vicinity
Honeydew Creek Bridge, Honeydew
Sacramento River Bridge, Isleton
Covered Bridge, Knights Ferry
Shafter Bridge, Lagunitas Vicinity
Carroll Overhead Bridge, Livermore Vicinity
San Antonio Creek Bridge, Lompoc Vicinity
Purdon Crossing Bridge, Nevada City Vicinity
San Francisco–Oakland Bay Bridge +, Oakland
Bidwell Bar Suspension Bridge +, Oroville Vicinity
Prospect Boulevard Bridge, Pasadena
Tobin Highway Bridge, Plumas
Union Pacific RR Bridge, Riverside
Victoria Bridge, Riverside
Tower Bridge, Sacramento
South Fork Trinity River Bridge, Salyer Vicinity
Pacheco Creek Bridge, San Felipe Vicinity
Alvord Lake Bridge +, San Francisco
Golden Gate Bridge +, San Francisco
San Roque Canyon Bridge, Santa Barbara

Tule River Hydroelectric Complex, Tule River Bridge, Springville Vicinity
Big Creek Bridge, Wawona Vicinity
Crane Creek Bridge, Wawona Vicinity
South Fork Merced River Bridge, Wawona Vicinity
Yosemite Creek Campground Bridge, Wawona Vicinity
Ahwahnee Bridge, Yosemite National Park
Bridal Veil Creek Bridge, Yosemite National Park
Bridalveil Falls Bridge I, Yosemite National Park
Bridalveil Falls Bridge II, Yosemite National Park
Cascade Creek Bridge, Yosemite National Park
El Capitan Bridge, Yosemite National Park
Happy Isles Bridge, Yosemite National Park
Old Happy Isles Bridge, Yosemite National Park
Pohono Bridge, Yosemite National Park
Sentinel Bridge, Yosemite National Park
South Fork Bridge, Yosemite National Park
Sugar Pine Bridge, Yosemite National Park
Tamarack Creek Bridge, Yosemite National Park
Tenaya Bridge, Yosemite National Park
Tuolumne Meadows Bridge, Yosemite National Park
Wawona Covered Bridge, Yosemite National Park
Wildcat Creek Bridge, Yosemite National Park
Yosemite Creek Bridge, Yosemite National Park

COLORADO
Nepesta Bridge, Boone Vicinity
Baseline Bridge, Brighton Vicinity
Four Mile Bridge, Buena Vista Vicinity
State Bridge, Del Norte Vicinity
Delta Bridge, Delta
Broadway Bridge, Denver
Cherry Creek RR Bridge, Denver
Colorado Historic Bridges Survey, Denver
Delgany Street RR Bridge, Denver
Fourteenth Street Viaduct, Denver
Nineteenth Street Bridge, Denver
South Platte Canyon Road Bridge, Denver Vicinity
Black Bridge, Grand Junction
Fifth Street Bridge, Grand Junction
Hotchkiss Bridge, Hotchkiss Vicinity
Keystone Bridge, Kassler Vicinity
Manzanola Bridge, Manzanola Vicinity
Hortense Bridge, Nathrop Vicinity
Uintah Railway: Whiskey Creek Trestle, Rangely Vicinity
Denver South Park & Pacific RR Truss Bridge, Romley
Four Mile Bridge, Steamboat Springs
Swink Bridge, Swink Vicinity
Commercial Street Bridge, Trinidad
Linden Ave. Bridge, Trinidad
Deansbury Bridge, Waterton Vicinity

CONNECTICUT
Patch Street Bridge, Danbury
New York,New Haven & Hartford RR: Niantic Bridge,

East Lyme
Niantic River Swing Bridge, East Lyme
Grasmere Ave. Bridge, Fairfield
Riverside Ave. Bridge, Greenwich
Bog Hollow Bridge, Kent
Chapel Street Swing Bridge, New Haven
Tomlinson Bridge, New Haven
Water Street Bridge, New Haven
New York, New Haven & Hartford RR: Groton Bridge, New London
Boardman's Bridge, New Milford
Lover's Leap Bridge, New Milford
Indian Leap Pedestrian Bridge, Norwich
Frost Bridge Road Bridge, Thomaston
Washington Ave. Lenticular Truss Bridge, Waterbury
Wilton Road Bridge (No. 727), Westport
Bridge Street Bridge, Windsor Locks

DELAWARE
Ashland Covered Bridge, Ashland
Smith's Bridge *, Granogue Vicinity
Rehoboth Ave. Bridge, Rehoboth
Augustine Bridge, Wilmington

DISTRICT OF COLUMBIA
Arlington Memorial Bridge, Washington
Boulder Bridge, Washington
Bridge over Boundary Channel, Washington
Grant Road Bridge, Washington
High Street Bridge *, Washington
Washington Old Military Road Bridge, Washington
Pinehurst Bridge, Washington
Rock Creek & Potomac Parkway Bridge near P St., Washington
Ross Drive Bridge, Washington
Shoreham Hill Bridge, Washington
Tidal Reservoir Inlet Bridge, Washington
Taft Memorial Bridge, Washington

FLORIDA
Seven Mile Bridge, Knight Key
Seddon Island Bridge, Tampa

GEORGIA
Woody Allen Road Bridge, Adairsville Vicinity
Marietta Road Bridge, Atlanta
Covered Bridge *, Atlanta Vicinity
Smith Bridge, Blairsville Vicinity
Broad River Highway Bridge, Carlton Vicinity
Dowhait's Bridge, Cartersville Vicinity
Lutens Bridge, Cash Vicinity
Alcovy Road Bridge, Covington
Fannin County Road 222 Bridge, Dial
Gordon County Road 220 Bridge, Fairmount Vicinity
Hart County Bridge, Hartwell Vicinity

Ocmulgee River Bridge, Hawinsville
Blackwell Bridge, Heardmont Vicinity
Headen Bridge, Hiawassee Vicinity
J. H. Millhollin Memorial Bridge, Jacksonville Vicinity
Curry Creek Bridge, Jefferson
Jefferson County Road 255 Bridge, Louisville
Gordon County Road No. 24 Bridge, New Town Vicinity
County Road 130 Bridge, Rising Fawn Vicinity
Second Ave. Bridge, Rome
Central of Georgia RR Bridges, Savannah
Central of Georgia RR: 1853 Brick Arch Viaduct *#, Savannah
Central of Georgia RR: 1860 Brick Arch Viaduct #, Savannah
Central of Georgia RR: Bay St. Viaduct, Savannah
Smith-McGee Bridge, Savannah
Georgia-Carolina Memorial Bridge, Savannah Vicinity
Haralson County Bridge, Tallapoosa Vicinity
Tifton Bridge, Tifton

HAWAII
Wailoa Bridge, Hilo

IDAHO
Fall River Bridge, Ashton
Bonner's Ferry Bridge, Bonner's Ferry
Hot Springs Bridge, Bruneau Valley Vicinity
Burton Road Bridge, Cambridge Vicinity
McCammon Overhead and River Crossing Bridge, Mccammon
Midvale Bridge, Midvale
East Dingle Bridge, Montpelier Vicinity
Murtaugh Bridge, Murtaugh Vicinity
Oldtown Bridge, Oldtown
Thatcher Bridge, Thatcher Vicinity
Twin Falls–Jerome Bridge, Twin Falls

ILLINOIS
Alton Bridge, Alton
MacArthur Bridge, Biggsville Vicinity
Cairo Bridge, Cairo
Suspension Bridge *, Carlyle
Chicago River Bascule Bridge: Michigan Ave., Chicago
Delray Bridge, Delray
Fall Creek Bridge Spanning Fall Creek Gorge *, Fall Creek Vicinity
Pecatonica River Bridge, Freeport Vicinity
Covered Wooden Bridge *, Homer Vicinity
Keithsburg Bridge, Keithsburg
London Mills Bridge, London Mills
Indian Ford Bridge, London Mills Vicinity
Eames Covered Bridge *, Oquawka Vicinity
Jack's Mill Covered Bridge *, Oquawka Vicinity
Bridge Spanning Mississippi River *, Rock Island
Rock Island (Government) Bridge, Rock Island
Vieley Bridge, Saunemin Vicinity

INDIANA
Deers Mill Covered Bridge, Alamo Vicinity
Mill Creek Bridge, Alton Vicinity
Laughery Creek Bridge, Aurora Vicinity
Jackson Covered Bridge, Bloomingdale
Brownsville Covered Bridge, Brownsville
Feederdam Bridge, Clay City Vicinity
Adams Mill Covered Bridge, Cutler
Dunlapsville Covered Bridge, Dunlapsville
Ceylon Covered Bridge, Geneva

Gosport Covered Bridge, Gosport
Hutsonville Bridge, Hutsonville
Brownsville Covered Bridge, Indianapolis
Vermont Covered Bridge, Kokomo
Mansfield Covered Bridge, Mansfield
Cumberland Covered Bridge, Mathews
Medora Covered Bridge, Medora
Leatherwood Station Covered Bridge, Montezuma
Hamilton County Bridge #218, Noblesville Vicinity
Gospel Street Bridge, Paoli
Tippecanoe River Bridge, Rochester Vicinity
Kennedy Covered Bridge *, Rushville Vicinity
Bells Ford Covered Bridge, Seymour
Narrows Covered Bridge, Turkey Run St. Park
Wabash River Bridge, Vera Cruz Vicinity
Busching Covered Bridge, Versailles
Delaware County Bridge No. 131, Yorktown

IOWA
Burlington Bridge, Burlington
MacArthur Bridge, Burlington Vicinity
Covered Bridge *, Carlisle Vicinity
Freeport Bridge, Decorah Vicinity
Eagle Point Bridge, Dubuque
Open Spandrel Bridge, Fort Dodge
Eureka Wrought Iron Bridge, Frankville Vicinity
Traer Street Bridge, Greene
Benton Street Bridge, Iowa City
Boyleston Bridge, Jackson Twp.
Lower Plymouth Rock Bridge, Kendallville Vicinity
Keokuk & Hamilton Bridge,
Keokuk Cottonville Bridge, Maquoketa Vicinity
South Third Ave. Bridge, Marshalltown
Rock Valley Bridge, Marshalltown Vicinity
Taylor Bridge, Mason City
Abraham Lincoln Memorial Bridge, Missouri Valley Vicinity
Mill Rock Bridge, Monmouth Township
Marsh Rainbow Arch Bridge, Newton
Nebraska City Bridge, Riverton Vicinity
Reinforced Concrete Arch Bridge, Rock Rapids Vicinity
Pacific Shortline Bridge, Sioux City
Sutliff's Ferry Bridge, Solon Vicinity

KANSAS
Parker Bridge, Coffeyville
Enterprise Parker Truss Bridge, Enterprise
Leavenworth Bridge, Leavenworth
Meriden Rock Creek Bridge, Meriden Vicinity
Covered Bridge *, Springdale Vicinity
Half-Mound Bridge, Valley Falls Vicinity

KENTUCKY
KY State Rt. 1032 Bridge, Berry
KY Rt. 49 Bridge, Bradfordsville
Covered Wooden Bridge *, Butler
KY Rt. 1754 Bridge, Chaplin Vicinity
Cincinnati Suspension Bridge +#, Covington Vicinity
Covered Wooden Bridge *, Cynthiana
Mitchell-Griggs Road Bridge, Dixon Vicinity
Red Bridge, Frankfort Vicinity
Starnes Bridge, Holbrook Vicinity
Sugar Creek Bridge, Hopkinsville Vicinity
Big Four Bridge, Louisville
KY Rt. 840 Bridge, Loyall
North Fork Bridge, Milford Vicinity
KY Rt. 762 Bridge, Owensboro Vicinity

U.S.23 Middle Bridge, Pikeville
Boldman Bridge, Pikeville Vicinity
KY Rt. 2014 Bridge, Pineville
Pine Street Bridge, Pineville
KY 708 Bridge, Tallege Vicinity
Williamsburg Bridge, Williamsburg
KY Rt. 478 Bridge, Williamsburg Vicinity
KY Rt. 228 Bridge, Wolf Creek

LOUISIANA
Krotz Springs Bridge, Krotz Springs
Bayou Teche Bridge, Ruth

MAINE
Lovejoy Bridge, Andover
Bailey Island Bridge +, Bailey Island
New Portland Wire Bridge, New Portland
Covered Bridge *, South Windham Vicinity

MARYLAND
Baltimore & Ohio RR: Carrollton Viaduct +#, Baltimore
Matthews Bridge, Baltimore
Cumberland & Penn. RR: Wills Creek Bridge, Cumberland Vicinity
C & O Canal: Lock 46 Roving Bridge *, Fort Frederick Vicinity
Jug Bridge *, Frederick Vicinity
Baltimore & Ohio RR: Waring Viaduct, Gaithersburg Vicinity
Cabin John Aqueduct Bridge *+#, Glen Echo
Casselman River Bridge *, Granstville Vicinity
Bridge Spanning Conocheague Creek *, Hagerstown Vicinity
Susquehanna River Bridge, Havre de Grace
Post Road Bridge, Havre de Grace Vicinity
Baltimore & Ohio RR: Patterson Viaduct *, Ilchester Vicinity
Covered Bridge *, Jerusalem Vicinity
Baltimore & Ohio RR: Long Bridge, Keelysville Vicinity
Keymar Bridge *, Keymar Vicinity
Baltimore & Ohio RR: Harpers Ferry Bridge Piers, Knoxville Vicinity
C & O Canal: White's Ferry Iron Bridge, Martinsburg Vicinity
Poffenberger Road Bridge, Middletown Vicinity
C & O Canal: Iron Bridge at Lock No. 68, Oldtown
Baltimore & Ohio RR: Thomas Viaduct *#, Relay
Old Mill Road Bridge, Rocky Ridge Vicinity
Baltimore & Ohio RR: Bollman Truss Bridge +, Savage
C & O Canal: Bridges, Aqueducts, etc., Seneca Vicinity
C & O Canal: Lock 25 Swing Bridge *, Seneca Vicinity
Burnside Bridge *, Sharpsburg Vicinity
Upper Bridge *, Sharpsburg Vicinity
Sharptown Bridge, Sharptown
Waverly Street Bridge, Westernport
Potomac Edison Company: C & O Canal Bridge, Williamsport
Salisbury Street Bridge, Williamsport

MASSACHUSETTS
Powow River Bridge, Amesbury
Essex-Merrimac Bridge, Amesbury/Newburyport
Boston & Maine RR: Clark Street Bridge, Belmont
Connecticut River RR: Fall River Viaduct, Bernardston
Boston & Maine RR: Charles River Bridges, Boston
Boston Public Garden: Suspension Bridge, Boston

Congress Street Bascule Bridge, Boston
Harvard Bridge, Boston
Longfellow Bridge, Boston
Northern Ave. Swing Bridge, Boston
NY, NH & Hartford RR: Fort Point Channel Bridge, Boston
Summer Street Retractile Bridge, Boston
Spring Street Bridge (Vine Rock Bridge), Boston/Dedham
Cape Cod Canal Lift Bridge, Buzzards Bay
Boston & Providence RR: Canton Viaduct, Canton
Bardwell's Ferry Bridge, Conway/Shelbourne
North Village Bridge, Dudley/Webster
French King Bridge, Erving/Gill
New Bedford–Fairhaven Middle Bridge, Fairhaven
(Lower) Rollstone Street Bridge, Fitchburg
Florida Bridge, Florida
Annisquam Bridge, Gloucester
Warner's Bridge *, Hamilton-Ipswich
Merrimac Bridge (Rocks Bridge), Haverhill
Boston & Worcester RR: Bogastow Brook Viaduct, Holliston
Holyoke Bridge, Holyoke
Choate Bridge, Ipswich
Atherton Bridge, Lancaster Vicinity
Ponakin Road Bridge, Lancaster Vicinity
Duck Bridge, Lawrence
Upper Pacific Mills Bridge, Lawrence
Tuttle Bridge, Lee
Aiken Street Bridge, Lowell
Covered Bridge *, Ludlow-Wilbraham
Middlesex Canal: Stone Bridge *, Medford
Covered Bridge *, Millville
Covered Bridge *, Montague City
Blackinton Bridge, N. Adams
Boston & Albany RR: Marion Street Bridge, Natick
New Bedford-Fairhaven Bridge, New Bedford
Sudbury River Aqueduct: Echo Bridge, Newton
Bay State Bridge, Northampton
Boston & Maine RR: Northampton
Lattice Truss Bridge, Northampton
Schell Memorial Bridge, Northfield
Bartlett's Bridge, Oxford
Nehemiah Jewett Bridge *, Pepperell
Woronoco Bridge, Russell
Hastings Bridge, Sterling
Butler Bridge, Stockbridge
Boston & Maine RR: Essex Street Bridge, Swampscott
Central Massachusetts RR: Linden St. Bridge, Waltham
Old Town Bridge *, Wayland
Boston & Albany RR: Kingsbury Street Bridge, Wellesley
Boston & Albany RR: Weston Road Bridge, Wellesley
Cheney Bridge, Wellesley
Fore River Bridge, Weymouth
Coleman Bridge, Windsor

MICHIGAN
Knaggs Bridge, Bancroft Vicinity
Dix Bascule Bridge, Detroit & Dearborn
West Knight Street Bridge, Eaton Rapids
Petrieville Road Bridge, Eaton Township
Dehmel Road Bridge, Frankenmuth Vicinity
Bridge Street Bridge, Grand Rapids
Pearl Street Bridge, Grand Rapids
Thirty-Sixth Street Bridge, Hamilton Vicinity
McCann Road Bridge, Hastings Vicinity

Washington Street Bridge, Hubbardston
Gull Street Bridge, Kalamazoo
Mill Street Bridge, Kalamazoo
Mosel Ave. Bridge, Kalamazoo
Cronk Road Bridge, Litchfield Vicinity
Monroe Street Bridge, Monroe
Lake Street Bridge, Muskegon
William S. Antisdale Memorial State Reward Bridge,
  Norton Shores
Bridge Street Bridge, Portland
Spruce Street Bridge, Sault Ste. Marie
Cheesman Road Bridge, St. Louis Vicinity

MINNESOTA
Duluth Aerial Lift Bridge, Duluth
Broadway Bridge, Minneapolis
Lake Street-Marshall Ave. Bridge, Minneapolis
Minnesota Veterans Home Complex: Steel Bridge *,
  Minneapolis
Steel Arch Bridge, Minneapolis
Bridge No. 4900, Rushford Vicinity
Sheridan Township Bridge, Sheridan Township
Jay Cooke State Park: Pedestrian Suspension Bridge,
  Silver Brook Twp.
Kern Truss Bridge, Skyline Vicinity
Smith Ave. High Bridge, St. Paul
Stone Highway Bridge *, Stillwater Vicinity
Winona Bridge, Winona

MISSISSIPPI
Bay Springs Bridge, Bay Springs Vicinity
Old Covered Bridge *, Steens Vicinity
Bridges of the Upper Tombigbee River Valley,
  Tombigbee Valley Vicinity
Confederate Ave. Bridge, Vicksburg
Keystone Bridge Company Bridge, Vicksburg
Ilinois Central Gulf RR Bridge, Vicksburg Vicinity

MISSOURI
Covered Bridge *, Allenville
Clear Creek Bridge, Ash Grove Vicinity
Leeper Ford Bridge, Ash Grove Vicinity
Bacon Bridge, Bacon Vicinity
Roberts Bluff Bridge, Blackwater Vicinity
Sunset Bridge, Bolivar Vicinity
Old Grade Road Bridge, Brashear Vicinity
Bollinger Covered Bridge & Mill, Burfordville
American Mill Bridge, Carthage Vicinity
Nineveh Bridge, Connelsville Vicinity
Dick's Mill Bridge, Cotton
Cedar Falls Road Bridge, Desloge Vicinity
Riddle Bridge, Dixon Vicinity
Current River Bridge, Doniphan Vicinity
Big Berger Creek Bridge, Etlah
East Fork Little Tarkio Bridge, Fairfax Vicinity
Steel Truss Bridge, Fairfield Vicinity
Surprise School Bridge, Gaines Vicinity
Grand River Bridge, Gentryville
Chicago & Alton Railway Bridge *, Glasgow
Sandy Creek Bridge *, Goldman Vicinity
William's Bend Bridge, Hermitage Vicinity
Noah's Arc Covered Bridge *, Hoover Vicinity
Noakes Bridge, Hopkins Vicinity
Lewis Mill Bridge, Jameson Vicinity
Hootentown Bridge, Jamesville Vicinity
Schneider's Ford Bridge, Jefferson
Jefferson Street Bridge, Jefferson City

Armour, Swift, Burlington Bridge, Kansas City
Rockhill Road Bridge, Kansas City
Windsor Harbor Road Bridge, Kimmswick
Gould Farm Bridge, Kingston Vicinity
Lock Springs Bridge, Lock Springs Vicinity
Trickum Road Bridge, Longwood Vicinity
Lime Kiln Road Bridge, Neosho Vicinity
Howard Ford Bridge, Nixa Vicinity
Osage River Bridge, Osceola
Osceola Bridge, Osceola
Paradise Road Bridge, Paradise Vicinity
Roscoe Bridge, Roscoe
Washington Ave. Bridge, Sedalia
James River Bridge, Springfield Vicinity
Defiance Road Bridge, St. Charles
Old St.Charles Bridge, St. Charles
Saxton Road Bridge, St. Joseph Vicinity
Eads Bridge *+#, St. Louis
Eighteenth Street Bridge, St. Louis
Kingshighway Viaduct, St. Louis
Tower Grove Park Bridges *, St. Louis
Twenty-first Street Bridge, St. Louis
Bellefontaine Bridge, St. Louis Vicinity
Waddell 'A' Truss Bridge, Trimble Vicinity
Middle Bridge, Warsaw
Warsaw Bridge, Warsaw
Bryan's Crossing Bridge, Warsaw Vicinity

MONTANA
Fish Creek Bridge, Alberton Vicinity
Dearborn River High Bridge, Augusta Vicinity
St. Mary River Bridge & Siphon, Babb Vicinity
Barber Bridge, Barber
Duck Creek Bridge, Billings Vicinity
Milk River Bridge at Coberg, Coberg
Coram Bridge, Coram Vicinity
Wolf Creek Bridge, Craig Vicinity
Deerfield Bridge, Danvers Vicinity
Sample's Crossing Bridge, Danvers Vicinity
Fort Benton Bridge, Fort Benton
Fromberg Bridge, Fromberg
Vandalia Bridge, Glasgow Vicinity
Tenth Street Bridge, Great Falls
Greycliff Bridge, Greycliff Vicinity
York Bridge, Helena Vicinity
Old Steel Bridge, Kalispell Vicinity
Pine Creek Bridge, Livingston Vicinity
Tongue River Bridge, Miles City
Fort Keogh Bridge, Miles City Vicinity
Kinsey Bridge, Miles City Vicinity
Paragon Bridge, Miles City Vicinity
Judith River Bridge, Moore Vicinity
Roundup Bridge, Roundup Vicinity
Snowden Lift Bridge, Snowden Vicinity
Springdale Bridge, Springdale
Calipso Bridge, Terry Vicinity
Dry Channel Bridge, Thompson Falls
Main Channel Bridge, Thompson Falls
Victor Bridge, Victor Vicinity
Avalanche Creek Bridge, West Glacier
Baring Creek Bridge, West Glacier
Belton Bridge, West Glacier
Divide Creek Bridge, West Glacier
Going-To-The-Sun Road System Bridges +, West
  Glacier
Logan Creek Bridge, West Glacier
Snyder Creek Bridge, West Glacier

St. Mary's Bridge, West Glacier

NEBRASKA
Abraham Lincoln Memorial Bridge, Blair Vicinity
Blair Crossing Bridge, Blair Vicinity
Nebraska City Bridge, Nebraska City Vicinity
Omaha Bridge, Omaha
Plattsmouth Bridge, Plattsmouth Vicinity
Rulo Bridge, Rulo

NEVADA
Muddy River Bridge, Glendale Junction
Riverside Bridge, Reno
East Verdi Bridge, Verdi
West Verdi Bridge, Verdi

NEW HAMPSHIRE
Osgood Bridge, Campton
Cornish-Windsor Covered Bridge +, Cornish
Durham Falls Bridge, Durham
Gleason Falls Bridge *, Hillsboro
Old Carr Bridge *, Hillsboro
Second New Hampshire Turnpike Bridge *, Hillsboro
Stone Bridge *, Hillsboro
Contoocook Covered Bridge *, Hopkinton
Covered Bridge *, Hopkinton Vicinity
Israel's River Bridge, Lancaster
Cohas Brook Bridge, Manchester
Notre Dame Bridge, Manchester
Covered Bridge *, Orford
Prescott Bridge, Raymond
Stone Bridge *, Stoddard Vicinity
Bellows Falls Arch Bridge, Walpole
Walpole-Westminster Bridge, Walpole

NEW JERSEY
Sun Gallery Bridge *, Atlantic City
Bayonne Bridge +, Bayonne
Finderne Ave. Bridge, Bridgewater Twp.
West Main Street Bridge, Clinton
South Broad Street Bridge, Elizabeth
Fink Through-Truss Bridge +, Hamden
Erie-Lackawanna RR Ferry Terminal: Slips & Bridges,
  Hoboken
Conrail Hack Lift Bridge, Kearny
Erie-Lackawanna RR Bridge, Kearny
Path Transit System Lift Bridge, Kearny
Maurice River Pratt Through-Truss Swing Bridge,
  Mauricetown
Grove Street Bridge, Montclair
Abbett Ave. Bridge, Morristown
Neshanic Station Lenticular Truss Bridge, Neshanic
  Station
Central RR of NJ: Newark Bay Lift Bridge, Newark
Erie Railway: New York Division Bridge 8.04, Newark
Jackson Street Bridge, Newark
Stony Brook Bridge *, Princeton
Covered Bridge *, Sergeantsville
Old Covered Bridge & Flood Gates, South Pemberton

NEW YORK
Hawk Street Viaduct, Albany
Whipple Cast & Wrought Iron Bowstring Truss Bridge,
  Albany
Prospect Street Bridge, Amsterdam
Rolling Hill Mill Road Bridge, Au Sable Forks
Lehigh Valley RR: Baltimore Through-Truss Bridge,

Batavia Vicinity
Tioronda Bridge, Beacon
Erie Railway: Allegany Division Bridge, Belfast Vicinity
Uncle Sam Bridge, Catskill
Erie Railway: Delaware Division Bridge 175.53,
  Deposit
Erie Railway: Oquaga Creek Bridge, Deposit
Hyde Hall Covered Bridge *, East Springfield
Hegeman-Hill Street Bridge, Easton
Erie Railway: Allegany Division, Bridge 367.33,
  Fillmore Vicinity
Fillmore Bridge, Fillmore Vicinity
Stone Bridge *, Fort Plain
Erie Canal Aqueduct *, Frankfort
Erie Canal Schoharie Creek Aqueduct #, Fort Hunter
New York & Mahopac RR: Bridge L-158, Goldens
  Bridge
Lordville Suspension Bridge, Hancock Vicinity
Old Croton Aqueduct: Quarry RR Bridge, Hastings-on-
  Hudson
Jones Beach Causeway Bridge No. 1, Hempstead
Sandy Hill Bridge, Hudson Falls
Boatlanding Bridge, Jamestown
Jay Covered Bridge, Jay
Upper Keeseville Bridge +, Keeseville
Erie Railway: Clear Creek Viaduct, Lawtons Vicinity
Leeds Bridge *, Leeds
Blood Road Bridge, Lyndonville Vicinity
Creager's Bridge, Lyons
Erie Railway: Sawyer Creek Bridge, Martinsville
Philipse Manor Station: Pedestrian Bridge, Mount
  Pleasant
Brooklyn Bridge +#, New York City
Central Park Bridges: Bow Bridge (Bridge 5) #,
  New York City
Central Park Bridges: Gothic Arch (Bridge 28) #,
  New York City
Central Park Bridges: Pine Bank Arch (Bridge No. 15)
  #, New York City
Central Park Bridges: Reservoir Bridge Southeast
  (Bridge 24) #, New York City
Central Park Bridges: Reservoir Bridge Southwest
  (Bridge 27) #, New York City
George Washington Bridge +, New York City
Manhattan Bridge, New York City
New York Connecting RR: Hell Gate Bridge, New
  York City
Queensboro Bridge, New York City
Verrazano Narrows Bridge, New York City
Washington Bridge, New York City
Williamsburg Bridge, New York City
Blenheim Covered Bridge *+#, North Blenheim
Long Island RR: Manhasset Bridge, North Hempstead
Lafayette-Spring Street Bridge, Ogdensburg
Old Croton Aqueduct: Sing Sing Kill Bridge, Ossining
Ouaquaga Bridge, Ouaquaga
Snyder Hollow Bridge, Phoenicia Vicinity
Erie Railway: Buffalo Division Bridge 361.66 (Portage
  Viaduct), Portageville Vicinity
Mid-Hudson Suspension Bridge, Poughkeepsie
Poughkeepsie Bridge, Poughkeepsie
Perrine's Bridge *, Rifton
Driving Park Ave. Bridge, Rochester
Main Street Bridge, Rochester
Rome Westernville Road Bridge, Rome
Baltimore & Ohio RR: Baltimore Skewed Through-Truss
  Bridge, Salamanca Vicinity

Cemetery Road Bridge, Salem
Erie Railway: Moodna Creek Viaduct, Salisbury Mills Vicinity
Old Bridge *, Saratoga Vicinity
Bayonne Bridge +, Staten Island
Locust Street Bridge, Waterloo
Court Street Bridge, Watertown
Quantuck Canal Bridge, Westhampton Beach
Saunders Street Bridge, Whitehall

NORTH CAROLINA
NC Rt. 2408 Bridge, Asheville Vicinity
NC Rt. 1336 Bridge, Burnsville Vicinity
Oconaluftee Bridge, Cherokee
NC Rt. 1417 Bridge, Danbury Vicinity
NC Rt. 1392 Bridge, Dillsboro Vicinity
Bridge No. 28, Durham Vicinity
NC Rt. 1761 Bridge, Eden Vicinity
McGirt's Bridge, Elizabethtown Vicinity
Person County Bridge No. 35, Hurdle Mills Vicinity
NC Rt. 1334 Bridge, Jamestown Vicinity
NC Rt. 1412 Bridge, Laboratory Vicinity
NC Rt. 1116 Bridge, Longview Vicinity
NC Rt. 126 Bridge, Marion Vicinity
Huffman Bridge, Morgantown Vicinity
Lake James Spillway Bridge, Nebo Vicinity
Boylan Ave. Bridge, Raleigh
NC Rt. 1314 Bridge, Relief Vicinity
Bridge No. 249, Southmont Vicinity
NC Rt. 1006 Bridge, Stoney Point Vicinity
NC Rt. 1852 Bridge, Tuxedo Vicinity

NORTH DAKOTA
Bismarck Bridge, Bismarck
Northern Pacific RR Overhead Bridge, Mandan
Mott Rainbow Arch Bridge, Mott

OHIO
'Forder' Pratt Through Truss Bridge, Antwerp Vicinity
Fourty-sixth Street Bridge, Ashtabula
Bladensburg Concrete Bowstring Bridge, Bladensburg Vicinity
Bridge on Old National Road *, Blaine
Bridge & Milestone *, Blaine Vicinity
Brecksville-Northfield High Level Bridge, Brecksville
Station Road Bridge, Brecksville
Bridge on Old National Road *, Cambridge
'S' Bridge *, Cambridge Vicinity
Bridge on Old National Road *, Cambridge Vicinity
Third Street Southeast Bridge, Canton
John Bright No. 1 Iron Bridge, Carroll Vicinity
John Bright No. 2 Covered Bridge, Carroll Vicinity
Chesapeake & Ohio RR Bridge, Cincinnati
Cincinnati Suspension Bridge *+#, Cincinnati
Covered Bridge *, Clarksville
Abbey Ave. Viaduct, Cleveland
B & O RR Bridge Number 464, Cleveland
Brookside Park Bridge, Cleveland
Carnegie-Lorain Bridge, Cleveland
Carter Road Lift Bridge, Cleveland
Center Street Bridge, Cleveland
Columbus Road Lift Bridge, Cleveland
Detroit Superior High Level Bridge, Cleveland
Old Detroit Street Bridge, Cleveland
Superior Ave. Viaduct, Cleveland
Wade Park Ave. Bridge *, Cleveland
Covered Bridge *, Collinsville Vicinity

Roberts Bridge *, Eaton Vicinity
Burrville Road Bridge, Fort Recovery Vicinity
New London Pike Covered Bridge *, Hamilton Vicinity
Old Covered Bridge *, Hopewell Vicinity
Bridge *, Limestone Vicinity
Benson Street Concrete Bowstring Bridge, Lockland
Smith Road Bowstring Arch Bridge, Lykens Vicinity
Harrison Road Camelback Through Truss Bridge, Miamitown
'S' Bridge *#, Middlebourne Vicinity
Bridge *, Middlebourne Vicinity
Millgrove Road Bridge, Morrow Vicinity
Covered Bridge *, Newton Falls
Covered Bridge *, North Lewisburg Vicinity
Main Street Bridge, Painesville
Main St. Parker Pony Truss Bridge, Poland
White Bowstring Arch Truss Bridge, Poland
Scioto Pennsylvania Through Truss Bridge, Portsmouth
Rocky River Bridge, Rocky River
First Street Reinforced Concrete Bridge, Roseville
Old Colerain Pennsylvania Through Truss Bridge, Ross Vicinity
Jaite Company RR Bridge, Sagamore Hills
Fosnaugh Truss Leg Bedstead Bridge, Stoutsville Vicinity
Abbott's Parker Through Truss Bridge, Tiffin Vicinity
Town Creek Truss-leg Bedstead Bridge, Van Wert Vicinity
Mahoning Ave. Pratt Double-Deck Bridge, Youngstown
Bridge *, Zanesville
Stone Bridge *, Zanesville
Y-Bridge, Zanesville
'S' Bridge *, Zanesville Vicinity
Bridge *, Zanesville Vicinity
Old Covered Bridge *, Zanesville Vicinity

OKLAHOMA
Bridge over Bird Creek, Avant
Hominy Creek Bridge, North Of Hominy
Pack Saddle Bridge, Roll Vicinity

OREGON
Albany Bridge, Albany
Columbia River Bridge at Astoria, Astoria
Latourell Creek Bridge, Bridal Veil
Shepperd's Dell Bridge, Bridal Veil
Depoe Bay Bridge, Cape Foulweather
Mosier Creek Bridge, Col. Riv. Hwy.
Columbia River Highway Bridges, Columbia River Vicinity
Coquille River Bridge, Coquille
Mills Creek Bridge, Dallas
Cape Creek Bridge, Florence Vicinity
Siuslaw River Bridge, Florence
Rogue River Bridge +, Gold Beach
Gold Hill Bridge over the Rogue River, Gold Hill
Rock Point Arch Bridge, Gold Hill
Rogue River (Caveman) Bridge, Grants Pass
Santiam River Bridge, Jefferson
Horse Creek Covered Bridge, Mckenzie Vicinity
Four Mile Bridge, Molalla Vicinity
Yaquina Bay Bridge, Newport
Coos Bay Bridge, North Bend
Crooked River (High) Bridge, Opal City
Williamette River Bridge, Oregon City
Oregon Coast Highway Bridges, Oregon Coast Vicinity
Broadway Bridge, Portland

Burlington Northern RR Bridge, Portland
Hawthorne Bridge, Portland
St. John's Bridge, Portland
Steel Bridge, Portland
Umpqua River Bridge, Reedsport
Jordan Covered Bridge, Scio Vicinity
Hayden Bridge, Springfield
Grave Creek Covered Bridge, Sunny Valley
Oregon Trunk RR Bridge, Terrebonne
Wilson River Bridge, Tillamook
Alsea Bay Bridge, Waldport
Moffett Creek Bridge, Warrendale
Dry Canyon Creek Bridge, White Salmon
Willamette River Bridges, Willamette River Vicinity
Winchester Bridge over the North Umpqua River, Winchester

PENNSYLVANIA
Haupt Truss Bridge, Altoona
Gross Covered Bridge, Beaver Springs Vicinity
Speicher Bridge, Bernville Vicinity
East Bloomsburg Bridge, Bloomsburg
Dunlaps Creek Bridge +, Brownsville
Erie Railway: Diverging French Creek Bridges, Cambridge Spr. Vicinity
Erie Railway: Parallel French Creek Bridges, Cambridge Spr. Vicinity
Catawissa Bridge, Catawissa Vicinity
Brevard Bridge, Chartiers Township
Christiana Borough Bridge, Christiana Borough
Cope's Bridge *, Copesville
Penna. RR: Erie RR Bridge, Corry Vicinity
Skew Arch Bridge *, Cresson Vicinity
Detters Mill Covered Bridge *, Detters Mill
Dingmans Ferry Bridge, Dingmans Ferry
Kuhn's Ford Bridge, East Berlin Vicinity
Emlenton Bridge, Emlenton
Tank Farm Road Bridge, Emmaus
Allentown Road Bridge, Franconia
Freemansburg-Steel City Bridge, Freemansburg
College Ave. Bridge, Greenville
Old Mill Road Bridge, Hellertown
Walnut Street Bridge, Hellertown
Snyder's Fording Covered Bridge *, Hunterstown Vicinity
Stevens Viaduct *, Iron Springs Vicinity
Hare's Hill Road Bridge, Kimberton
Delaware & Hudson Canal: Delaware Aqueduct +#, Lackawaxen
Erie Railway: Delaware Division Bridge 110.54, Lackawaxen
Erie Railway: Starrucca Viaduct +, Lanesboro
Erie Railway: Delaware Division Bridge 190.13, Lanesboro
Erie Railway: Cascade Bridge Site, Lanesboro Vicinity
St. Anthony Street Bridge, Lewisburg
Stoneroads Mill Bridge *, Maple Grove Vicinity
Johnson's Mill Bridge *, Marietta Vicinity
Mead Ave. Bridge, Meadville
Erie Railway: Delaware River Bridge, Millrift
New York & Erie Railway: Kinzua Viaduct +, Kushequa
Reading-Halls Station Bridge, Muncy Vicinity
DL&W RR: Tunkhannock Viaduct +, Nicholson
West Marshall Street Bridge, Norristown
Erie Railway: Meadville Division Bridge 33.14, Oil City
Pennsylvania RR: Allegheny River Bridge, Oil City
Callowhill Street Bridge, Philadelphia

Chestnut Street Bridge *, Philadelphia
Covered Bridge *, Philadelphia
Falls Bridge, Philadelphia
Girard Ave. Bridge *, Philadelphia
Pennsylvania RR: Brick Viaduct, Philadelphia
Pennsylvania RR: Mantua Junction Viaduct, Philadelphia
Pennypack (Frankford Ave.) Creek Bridge *+, Philadelphia
Philadelphia & Reading RR: Schuylkill River Viaduct, Philadelphia
Strawberry Mansion Bridge *, Philadelphia
Upper Ferry Bridge *, Philadelphia
Leaman Rifle Works Bridge *, Pinetown
Brady Street Bridge *, Pittsburgh
Manchester Bridge, Pittsburgh
North Side Point Bridge, Pittsburgh
Point Bridge *, Pittsburgh
Smithfield Street Bridge +#, Pittsburgh
Washington Crossing Bridge *, Pittsburgh
West End-North Ave. Bridge, Pittsburgh Vicinity
Philadelphia & Reading RR: Skew Arch Bridge, Reading
Philadelphia & Reading RR: Walnut Street Bridge, Reading
Philadelphia & Reading RR: Peacock's Lock Viaduct, Reading Vicinity
Delaware River Bridge, Riegelsville
Covered Bridge *, Ruff Creek Vicinity
Scarlets Mill Bridge, Scarlets Mill
D L & W RR: Scranton Yards Bridge 60, Scranton
D L & W RR: Scranton Yards Cedar Ave. Bridge, Scranton
D L & W RR: Scranton Yards Washington Ave. Bridge, Scranton
Sanderson Ave. Bridge, Scranton
Sewickley Bridge, Sewickley
Erie Railway: Shohola Creek Bridge, Shohola
Smithton Bridge, Smithton
Griesemer Mill Covered Bridge *, Spangsville Vicinity
Stewartstown RR Bridge, Stewartstown
Bridge Street Bridge, Union City
Henszey's Wrought-Iron Arch Bridge, Wanamakers
Washingtonville Bridge, Washingtonville Vicinity
Waterford Covered Bridge, Waterford
Bells Mill Bridge *, West Newton Vicinity
Bartram's Covered Bridge *, Whitehorse Vicinity
South Street Bridge, Wilkes-Barre
Memorial Ave. Bridge, Williamsport

RHODE ISLAND
Albion Bridge, Cumberland
Albion Trench Bridge, Cumberland
Charles Street Bridge, Providence
Exchange Bridge, Providence
Francis Street Bridge, Providence
Gaspee Street Bridge, Providence
Promenade Street Bridge, Providence
Union Station Viaduct, Providence
Woonasquatucket Bridge, Providence

SOUTH CAROLINA
Saluda 1 Bridge, Chapples Ferry Vicinity
Gervais Street Bridge, Columbia
Chapman's Bridge *, Gowensville Vicinity
Saluda County Bridge No. 4, Jones Crossroads Vicinit
Lower Gassaway Bridge *, Norris Vicinity
Edgefield County Bridge No. 3, Pleasant Lane Vicinity

Sanders Ferry Bridge, Savannah River
Seaboard Coast Line RR Bridge, Savannah River
Poinsett Bridge, Tigerville Vicinity

SOUTH DAKOTA
Sioux City Bridge, Sioux City

TENNESSEE
Dentville Road Bridge, Benton Vicinity
Riverside Bridge, Bluff City Vicinity
Buena Vista Ford Bridge, Carthage Vicinity
Walnut Street Bridge, Chattanooga
John T. Cunningham Memorial Bridge, Clarksville
Hobbs Bridge, Coldwater Vicinity
Boulevard Bridge, Dechard
Elizabethton Covered Bridge *, Elizabethton
McPherson Bridge, Eureka
Coldwater Bridge, Fayetteville Vicinity
Stone Bridge at Bowling Green *, Gallatin Vicinity
Weaver Road Bridge, Kingston Vicinity
Asylum Bridge, Knoxville
Lenox Bridge, Lenox Vicinity
Memphis Bridge +, Memphis
Newsom's Mill Bridge, Nashville Vicinity
Massengill Bridge, Norris Vicinity
Moore Road Bridge, Unionville Vicinity
Gibson Bridge, Watauga Flats Vicinity

TEXAS
Bird Pond Road Bridge, College Station Vicinity
San Antonio Acquia: Piedras Creek (Espada Aqueduct)
    + #, San Antonio
Waco Suspension Bridge, Waco

UTAH
Benson Bridge, Benson
San Rafael Bridge, Glen Canyon Vicinity
Henefer Bridge, Henefer
Virgin River Bridge, Hurricane Vicinity
Jensen Bridge, Jensen
Marysvale Bridge, Marysvale
Union Pacific RR: Gateway Bridge, Ogden
Fairmont Bridge, Ogden Canyon
Southern Pacific RR: Ogden-Lucin Cutoff Trestle,
    Promontory
Telluride Power Company: Provo River Bridge, Orem
    Vicinity
Denver & Rio Grande Western RR: Provo River Bridge,
    Orem Vicinity (Olmstead)
Zion-Mount Carmel Highway: Pine Creek Bridge,
    Springdale Vicinity
Zion-Mount Carmel Highway: Virgin River Bridge,
    Springdale Vicinity
Standard Mines: Timber Trestle, Standardville Vicinity
Uintah Bridge, Uintah

VERMONT
Northfield Parker Truss Bridge, Northfield
E. & T. Fairbanks & Co., Two-Story Covered Bridge,
    St. Johnsbury
Mount Vernon Street Bridge, St. Johnsbury
Elm Street Bridge, Woodstock

VIRGINIA
Mount Vernon Memorial Hwy: Alexandria Ave. Bridge,
    Alexandria Vicinity

Mount Vernon Memorial Hwy: Hunting Creek Bridge,
    Alexandria Vicinity
Arlington Memorial Bridge, Arlington
Roaring Run Bowstring Truss Bridge, Bedford Vicinity
Daphna Creek Pratt Truss Bridge, Broadway
Thacher Truss Bridge, Broadway
James River Suspension Bridge, Buchanan
Cartersville Bridge, Cartersville Vicinity
Humpback Covered Bridge, Covington Vicinity
South River Pratt Through-Truss Bridge, Crimora
    Vicinity
Covered Bridge *, Cumberland Vicinity
North Branch Quantico Creek Bridge, Dumfries Vicinity
South Branch Quantico Creek Bridge, Dumfries
    Vicinity
J. H. C. Mann Bridge, East Lexington Vicinity
Marysville Covered Bridge, Gladys Vicinity
Covered Bridge, Lexington Vicinity
James River & Kanawha Canal Bridge *, Lynchburg
Sixth Street Bridge, Lynchburg
Mount Vernon Mem. Hwy: Little Hunting Creek Bridge,
    Mount Vernon Vicinity
Bridge No. 6120, Owensville Vicinity
Marshall Street Viaduct, Richmond
Seventh Street Bridge, Richmond
Valley RR: Folly Mills Creek Viaduct, Staunton Vicinity
Middle River Pratt Through-Truss Bridge, Weyers Cave
    Vicinity

WASHINGTON
West Wishkah Bridge, Aberdeen Vicinity
Grays River Covered Bridge, Grays River Vicinity
'F' Street Bridge, Palouse
McClure Bridge, Palouse Vicinity
Pasco-Kennewick Bridge, Pasco
Grant Ave. Bridge, Prosser
Washington Street Bridge, Spokane
Chow Chow Suspension Bridge, Toholah Vicinity
Yakima Valley Trans. Co.:Naches River Bridge, Yakima
    Vicinity

WEST VIRGINIA
Barrickville Covered Bridge, Barrickville
Baltimore & Ohio RR: Benwood Bridge, Benwood
Bridgeport Lamp Chimney Co.: Simpson Creek Bridge,
    Bridgeport
Baltimore & Ohio RR: Fairmont Bridge, Fairmont
    Vicinity
New River Gorge Bridge, Fayetteville Vicinity
Baltimore & Ohio RR: Stone Bridges 1 & 2, Gassaway
    Vicinity
Northwestern RR: Grafton Bridge, Grafton
Baltimore & Ohio RR: Cacapon River Viaduct, Great
    Cacapon
Baltimore & Ohio RR: Bollman Bridge, Harpers Ferry
Milton Covered Bridge, Milton
Baltimore & Ohio RR: Parkersburg Bridge, Parkersburg
Staats Mill Covered Bridge, Ripley Vicinity
Baltimore & Ohio RR: Rowelsburg Bridge, Rowelsburg
    Vicinity
Baltimore & Ohio RR: Tray Run Viaduct, Rowelsburg
    Vicinity
Bridge on Old National Trail *, Wheeling
Bridgeport Bridge, Wheeling
Wheeling Suspension Bridge + #, Wheeling
Williamstown-Marietta Bridge, Williamstown

WISCONSIN
Range Line Road Bridge, Ackley
Beaver Dam Road Bridge, Addison
Sprague Bridge, Armenia & Necedah
Yellow River Bridge, Arthur
Manchester Street Bridge, Baraboo
Turtleville Iron Bridge, Beloit Vicinity
Coltman Bridge, Benton
Bear Creek Bridge, Buena Vista
White River Bridge, Burlington
Covered Bridge *, Cedarburg Vicinity
Chester Bridge, Chester Township
Bridge of Pines, Chippewa Falls
Spring Street Bridge, Chippewa Falls
Fountain Island Bridge, Fond du Lac
Branch River Bridge, Franklin
Kennan-Jump River Bridge, Kennan
McGilvray Road Bridge No. 1, La Crosse Vicinity
McGilvray Road Bridge No. 2, La Crosse Vicinity
McGilvray Road Bridge No. 3, La Crosse Vicinity
McGilvray Road Bridge No. 4, La Crosse Vicinity
McGilvray Road Bridge No. 6, La Crosse Vicinity
Bridge No. 18, Lafarge
State Highway Bridge No. 16, Lafarge Vicinity
Sock Road Bridge, Lowell
Marathon City Bridge, Marathon
Chicago & North Western Railway Bridge No. 344,
    Merrimack Vicinity
Grand Ave. Viaduct, Milwaukee
Highland Boulevard Viaduct, Milwaukee
Lake Park Brick Arch Bridge, Milwaukee
Lake Park Lions Bridge, Milwaukee
North Ave. Viaduct, Milwaukee
Lynch Bridge, Neillsville
New London Bridge, New London Vicinity
Eau Claire Dells Bridge, Plover
Sauk Creek Bridge, Port Washington
Prescott Bridge, Prescott Vicinity
Horlick Drive Bridge, Racine
West Sixth Street Bridge, Racine
Cedar Street Bridge, River Falls
Cunningham Lane Bridge, Rockbridge
Mill Bridge, Scandinavia
Eau Claire River Bridge, Schofield
Wagon Trail Road Bridge, Spring Valley
Palm Tree Road Bridge, St. Cloud
Green Bay Road Bridge, Thiensville
Chicago & North Western Railway Bridge No. 128,
    Tiffany
Wisconsin-Michigan RR Bridge, Wagner
Hemlock Bridge, Warner Twp.
Milwaukee Street Bridge, Watertown
Tivoli Island Bridge, Watertown
Ferndale Road Bridge, Wausaukee Vicinity
Wrightstown Bridge, Wrightstown

WYOMING
Arvada Bridge, Arvada Vicinity
Powder River Bridge, Arvada Vicinity
Laramie River Bridge, Bosler Vicinity
New Fork River Bridge, Boulder Vicinity
Peloux Bridge, Buffalo Vicinity
Bessemer Bend Bridge, Casper Vicinity
Hayden Arch Bridge, Cody Vicinity
Green River Bridge, Daniel Vicinity
Big Wind River Bridge, Dubois Vicinity
Medicine Bow Bridge, Elk Mountain Vicinity

Butler Bridge, Encampment Vicinity
Wind River Bridge, Ethete Vicinity
Green River Bridge, Fontenelle Vicinity
Black's Fork Bridge, Fort Badger Vicinity
Fort Laramie Bowstring Arch Truss Bridge, Fort
    Laramie Vicinity
Big Island Bridge, Green River Vicinity
Missouri River Bridge, Hulett Vicinity
South Fork Powder River Bridge, Kaycee Vicinity
Powder River Bridge, Leiter Vicinity
Shoshone River Bridge, Lovell Vicinity
Tongue River Bridge, Monarch Vicinity
Wind River Diverson Dam Bridge, Morton Vicinity
Cheyenne River Bridge, Riverview Vicinity
Pick Bridge, Saratoga Vicinity
Shell Creek Bridge, Shell Vicinity
Big Goose Creek Bridge, Sheridan Vicinity
Irigary Bridge, Sussex Vicinity
Owl Creek Bridge, Thermopolis Vicinity
Four Mile Bridge, Thermopolis Vicinity
Laramie River Bridge, Wheatland Vicinity
Chittenden River Bridge, Yellowstone National Park
Crawfish Creek Bridge, Yellowstone National Park
Cub Creek Bridge, Yellowstone National Park
FHWA Creek Bridge, Yellowstone National Park
Firehole River Bridge, Yellowstone National Park
Fishing Bridge, Yellowstone National Park
Gibbon River Bridge No. 1, Yellowstone National Park
Gibbon River Bridge No. 2, Yellowstone National Park
Golden Gate Viaduct, Yellowstone National Park
Indian Creek Bridge, Yellowstone National Park
Isa Lake Bridge, Yellowstone National Park
Lava Creek Bridge, Yellowstone National Park
Lemar River Bridge, Yellowstone National Park
Mammoth High Bridge, Yellowstone National Park
Midway Bridge, Yellowstone National Park
Nez Perce Bridge, Yellowstone National Park
Ojo Caliente Bridge, Yellowstone National Park
Otter Creek Bridge, Yellowstone National Park
Otter Creek Bridge No. 2, Yellowstone National Park
Pebble Creek Bridge, Yellowstone National Park
Pelican Creek Bridge, Yellowstone National Park
RWC Creek Bridge, Yellowstone National Park
Sedge Creek Bridge, Yellowstone National Park
Seven Mile Bridge, Yellowstone National Park
Soda Butte Creek Bridge No. 1, Yellowstone National
    Park
Soda Butte Creek Bridge No.2, Yellowstone National
    Park
TLF Creek Bridge, Yellowstone National Park
Tower Creek Bridge, Yellowstone National Park
Tower Suspension Bridge, Yellowstone National Park
Yellowstone Roads and Bridges, Yellowstone National
    Park

Bibliography

This bibliography includes books, pamphlets, and articles used by the author in developing the time line that appears throughout the book, the captions for the pictures, and the introductory essay.

Boller, Alfred Pancoast. *Practical Treatise on the Construction of Iron Highway Bridges.* J. Wiley & Sons. New York. 1885.

Breman, Paul (Editor). *Architecture, Catalogue 9: Bridges.* B. Weinreb, Ltd. London. 1965.

Burnell, George R., Clark, William T. (Editors) *Supplement to the Theory, Practice and Architecture of Bridges.* J. Weale. London. 1852-3.

Campin, Francis. *A Treatise on the Application of Iron to the Construction of Bridges, Girders, Roofs and Other Works.* Lockwood & Co. London. 1871.

Campin, Francis. *Iron and Steel Bridge and Viaducts.* Crosby Lockwood. London. 1898.

Chamberlin, William P. *Historic Bridges Criteria for Decision Making.* National Cooperative Highway Research Program, Synthesis of Highway Practice 101. Transportation Research Board, National Research Council. Washington, D.C. 1983.

Condit, Carl W. *American Building Art 19th & 20th Century.* 2 volumes. Oxford University Press. New York. 1961.

Darnell, Victor C. *A Directory of American Bridge-Building Companies: 1840 1900.* Occasional Publication No.4. Society for Industrial Archeology. Washington, D.C. 1984.

Darnell, Victor C. "The Haupt Iron Bridge on the Pennsylvania Railroad." *IA: The Journal of the Society for Industrial Archeology.* Vol. 14, No. 2. Washington, D.C. 1988.

DeLony, Eric N. "Conflict Between Structurally Deficient & Historically Significant Bridges." *Final Report: Maintenance, Repair and Rehabilitation of Bridges.* Vol.39. IABSE Symposium, Washington, D.C. 1982.

DeLony, Eric N. "HAER's Historic Bridge Program." *IA: The Journal of the Society for Industrial Archeology.* Vol. 15, No. 2. Washington, D.C. 1989.

Du Bois, A. Jay. *The Stresses in Framed Structures.* J. Wiley & Sons. New York. 1900.

DuFour, Frank O. *Bridge Engineering: Roof Trusses.* American School of Correspondence. Chicago. 1913.

Duggan, George. *Specimens of Stone, Iron and Wood Bridges, Viaducts, Tunnels and Culverts, &c. of the United States Railroads.* Various publishers. New York. 1850.

Edwards, Llewellyn Nathaniel. *A Record of History & Evolution of Early American Bridges.* University of Maine Press. Orono. 1959.

Ellet, Charles, Jr. *Report on a Suspension Bridge Across the Potomac, for Rail Road & Common Travel.* John C. Clark, printer. Philadelphia. 1852.

Elton, Julia. *Catalogue Number 2: Engineering & Engineers, Bridges - Rivers & Canals, Railways & The Steam Engine, Electric Telegraph.* Elton Engineering Books. London. 1988.

Elton, Julia. *Catalogue Number 4: Suspension Bridges: Their History & Development with Some Related Works.* Elton Engineering Books. London. 1989.

Fairbairn, William. *An Account of the Construction of the Britannia & Conway Tubular Bridges.* J. Weale and Longman, Brown, Green & Longmans. London. 1849.

Great Britain. *Report of the Commissioners Appointed to Inquire into the Application of Iron to Railway Structures.* William Clowes & Sons. London. 1849.

Haupt, Herman. *General Theory of Bridge Construction.* D. Appleton & Co. New York. 1866.

Hool, George A., and Kinne, W.S. *Moveable and Long Span Steel Bridges.* McGraw-Hill. New York. 1923.

Hopkins, H.J. *A Span of Bridges: An Illustrated History.* David & Charles. Newton Abbot, Devon, England. 1970.

Hosking, William. *Essay and Treatises on the Practice & Architecture of Bridges.* London. 1843.

Hovey, Otis Ellis. *Moveable Bridges.* Vol. I, superstructure; Vol. II, machinery. J. Wiley & Sons. New York. 1926, 1927.

Humber, William. *A Practical Treatise on Cast & Wrought Iron Bridges & Girders.* 2 volumes. 1857.

Hutchinson, Edward. *Girder Making and the Practice of Bridge Building in Wrought Iron.* E. & F. N. Spon. London. 1879.

Jackson, Donald C. *Great American Bridges & Dams.* Great American Place Series. The Preservation Press. Washington, D.C. 1988.

James, John G. "The Evolution of Iron Bridge Trusses to 1850." *The Newcomen Society for the Study of the History of Engineering & Technology Transactions.* Vol. 52. 1980–81. London. 1982.

Johnson, John B., Bryan, C.W., and Turneaure, F. E. *The Theory & Practice of Modern Framed Structures.* J. Wiley & Sons. New York. 1893.

Kemp, Emory L. "Thomas Paine & His Pontifical Matters." *The Newcomen Society for the Study of the History of Engineering & Technology Transactions.* Vol. 49. 1977–78. London. 1979.

Ketchum, Milo Smith. *The Design of Highway Bridges and the Calculation of Stresses in Bridge Trusses.* The Engineering News Publishing Co. New York. 1880.

Ketchum, Milo Smith. *The Design of Highway Bridges of Steel, Timber and Concrete.* McGraw-Hill. New York. 1920.

Kunz, F. C. *Design of Steel Bridges.* McGraw-Hill. New York. 1915.

Luten, Daniel Benjamin. *Reinforced Concrete Bridges.* Hollenbeck Press. Indianapolis. 1917.

Mahan, D. H. *A Treatise on Civil Engineering.* J. Wiley & Sons. New York. 1873.

Merrill, Bvt. Col. William E. *Iron Truss Bridges for Railroads.* D. Van Nostrand. New York. 1870.

Merriman, Mansfield, and Jacoby, Henry S. *A Text-Book on Roofs & Bridges.* Part 1, stresses in simple trusses. Part 2, graphic statics. Part 3, bridge design. J. Wiley & Sons. New York. 1904.

Peters, Tom F. *Transitions in Engineering.* Birkhauser Verlag. Basel, Switzerland, and Boston. 1987.

Plowden, David. *Bridges: The Spans of North America.* The Viking Press. New York. 1974.

Pope, Thomas. *A Treatise on Bridge Architecture.* Printed for the author by A. Niven. 1811.

Roebling, John A. *Long & Short Span Railway Bridges* D. Van Nostrand. New York. 1869.

Sayenga, Donald. *Ellet & Roebling.* American Canal & Transportation Center.

York, Pa. 1983.

Schodek, Daniel L. *Landmarks in American Civil Engineering.* MIT Press. Cambridge, Mass., and London. 1987.

Secretary of Transportation. *Highway Bridge Replacement & Rehabilitation Program 1991 of the Secretary of Transportation to the United States Congress.* Washington, D.C. 1991.

Smith, H. Shirley. *The World's Great Bridges.* Harper & Row. New York. 1965.

Stephens, John H. *Towers, Bridges and Other Structures.* Sterling Publishing Co. New York. 1976.

Thorpe, William Henry. *The Anatomy of Bridgework.* E. & F. N. Spon. London and New York. 1906.

Timoshenko, Stephen P. *History of Strength of Materials.* McGraw-Hill. New York. 1953.

Town, Ithiel. *A Description of Ithiel Town's Improvement in the Construction of Wood and Iron Bridges:* Printed by S. Converse. New Haven. 1821.

Tyrrell, Henry Grattan. *Concrete Bridges & Culverts for Both Railroads & Highways.* M.C. Clark. Chicago and New York. 1909.

Tyrrell, Henry Grattan. *History of Bridge Engineering.* Published by the author. Chicago. 1911.

Vogel, Robert M. *Roebling's Delaware & Hudson Canal Aqueducts.* Smithsonian Studies in History and Technology, No. 10. Smithsonian Institution Press. Washington, D.C. 1971.

Vogel, Robert M. *The Engineering Contributions of Wendel Bollman.* Contributions from the Museum of History & Technology, Paper 36. Smithsonian Institution Press. Washington, D.C. 1964.

Vose, George Leonard. *Bridge Disasters in America: The Cause & Remedy.* Lee and Shepard. Boston. 1887.

Waddell, John Alexander Low. *Bridge Engineering.* 2 volumes. J. Wiley & Sons. New York. 1916.

Waddell, John Alexander Low. *Economics of Bridgework.* J. Wiley & Sons. New York. 1921.

Watson, Wilbur J. *Bridge Architecture.* W. Helburn. New York. 1927.

Weale, John (Editor). *Theory, Practice & Architecture of Bridges of Stone, Iron, Timber, and Wire.* Architectural Library. London. 1843.

Weitzman, David. *Traces of the Past: A Field Guide to Industrial Archeology.* Charles Scribner's Sons. New York. 1980.

Whipple, Squire. *An Elementary & Practical Treatise on Bridge Building.* D. Van Nostrand. New York. 1873.

Whipple, Squire. *A Work on Bridge Building.* Published by author. New York. 1847.

White, Joseph & von Bernewitz, M.W. *The Bridges of Pittsburgh.* Cramer Printing & Publishing. Pittsburgh. 1928.

Whitney, Charles S. *Bridges: A Study of Their Art, Science and Evolution.* William Edwin Rudge. New York. 1929.

Wood, De Volson. *Treatise on the Theory of the Construction of Bridges and Roofs.* J. Wiley & Sons. New York. 1876.

Woodward, C. M. *A History of the St. Louis Bridge.* G.I. Jones & Co. St. Louis. 1881.

Index

The name of the photographer and the date the photograph was taken follow each bridge listing (line artist).*

Alsea Bay Bridge, Jet Lowe, 1990, p. 129
Alvord Lake Bridge, Jet Lowe, 1984, p. 95
Bailey Island Bridge, Jet Lowe, 1984, pp. 122–123
Barrickville Covered Bridge,
     William Barrett, 1978, p. 26
     Frederick Love, 1973*, p. 27
Bayonne Bridge, Jet Lowe, 1985, pp. 136–137
Bellows Falls Arch Bridge,
     Eric DeLony, 1978, p. 107
Blenheim Covered Bridge, Nelson E. Baldwin, 1936
     A.K. Mosley, 1936*, p. 29
Bollman Truss Bridge, Eric DeLony, 1990
     Brian Bartholomew, 1987*, p. 53
Bow Bridge, Jet Lowe, 1984, pp. 46–47
Bridge 28 (Gothic Arch), Jet Lowe, 1984, p. 50
Bridgeport Covered Bridge, Jet Lowe, 1983, p. 32
Brooklyn Bridge, Jet Lowe, 1987
     Eric DeLony, 1985, pp. 92–93
Brownsville Bridge, Jack Boucher, 1974
     Mark Mattox, 1971*, pp. 22–23
Cabin John Aqueduct, Jet Lowe, 1985, pp. 20–21
Canton Viaduct, Jet Lowe, 1982, p. 12
Cape Creek Bridge, Jet Lowe, 1990, p. 128
Casselman Bridge, A.S. Burns, 1933, pp. 6–7
Center Street Bridge, Jet Lowe, 1979, pp. 104–105
Central of Georgia RR: Brick Arch Viaducts,
     Jack Boucher, 1975, pp. 18, 19
     John G. Albers, 1975*, pp. 18–19
Choate Bridge, Jet Lowe, 1983, pp. 4–5
Cincinnati Suspension Bridge, Jack Boucher, 1982
     Eric DeLony, 1973, pp. 38–39
Coos Bay (McCullough Memorial) Bridge,
     Jet Lowe, 1990, pp. 134–135
Cornish-Windsor Covered Bridge,
     Jet Lowe, 1984, p. 33
Delaware Aqueduct, Jack Boucher, 1971, p. 34
     David Plowden, 1969, p. 34
     Eric DeLony, 1986, p. 35
     Scott Barber, 1988*, pp. 34–35
Dunlaps Creek Bridge, Jet Lowe, 1984, p. 13
Eads Bridge, Jet Lowe, 1983, p. 65
Elm Street Bridge, Eric DeLony, 1976, pp. 62–63
Eureka Wrought Iron Bridge.
     J Ceronie, 1983, pp. 66–67
Fink Through-Truss Bridge, Jack Boucher, 1974

Lori Allen, 1985*, p. 44
Fort Laramie Bowstring Arch-Truss Bridge,
     Jack Boucher, 1974, pp. 70–71
Freeport Bridge, Clayton B. Fraser, 1985*, p. 78
George Washington Bridge,
     Jet Lowe, 1986, 1991, pp. 138–139
Golden Gate Bridge, Jet Lowe, 1984, p. 143
Hare's Hill Road Bridge, Joseph Elliot, 1991
     Monika Korsos, 1991*, p. 54
Haupt Truss Bridge.
     Christine Theodoropoulos, 1992*, p. 28
Hayden Bridge, Jet Lowe, 1990
     Todd Croteau, 1990*, pp. 80–81
Hell Gate Bridge, Jack Boucher, 1978, p. 120
Henszey's Wrought Iron Arch Bridge,
     Jet Lowe, 1982, p. 55
     Wayne Chang, 1991*, p. 55
Humpback Covered Bridge, Jack Boucher, 1971
     Charles King, 1970*, pp. 30–31
John Bright No. 1 Iron Bridge,
     Louise Taft Cawood, 1986, pp. 84–85
Kinzua Viaduct, Jack Boucher, 1972, pp. 102–103
Laughery Creek Bridge, Jack Boucher, 1974
     M. Turner* (no date), pp. 74–75
Lover's Leap Bridge, Jet Lowe, 1984, pp. 88–89
Lower Plymouth Rock Bridge,
     Clayton B. Fraser, 1985, pp. 76–77
Manchester Bridge, Charles W. Shane, 1970,
     City of Pittsburgh, Dept. of Public Works,
     Bureau of Engineering, 1916, p. 118
Manhattan Bridge, David Sharpe, 1977, p. 110
Memphis Bridge, Clayton B. Fraser, 1985, pp. 98–99
Million Dollar Bridge, Jet Lowe, 1984, p. 112–113
New Portland Wire Bridge,
     Jet Lowe, 1984, pp. 40–41
Northern New Jersey Bridges,
     Jack Boucher, 1978, p. 121
Ogden-Lucin Cutoff Trestle, Jack Boucher, 1971
     Robert McNair, 1971*, p. 106
Old Mill Road Bridge, Jet Lowe, 1982
     Coy Burney, 1986*, pp. 58–59
Ouaquaga Bridge, Martin Stupich, 1987
     Charissa Wang, 1987*, pp. 86–87
Pacific Short Line Bridge,
     Hans Meussig, Marie Neubauer, and
     Richard Anderson Jr., 1981*, p. 101
Pine Bank Arch, Jet Lowe, 1984, p. 47
Poinsett Bridge, Jack Boucher, 1986, pp. 8–9
Poughkeepsie Bridge, Jack Boucher, 1978, p. 94

Pulaski Skyway, Eric DeLony, 1978, pp. 140–141
Queensboro Bridge, Jack Boucher, 1970, p. 111
Reading-Halls Station Bridge,
     Jet Lowe, 1984, pp. 24, 25
     Richard Anderson Jr., 1987*, p. 24
Reservoir Bridge Southeast, Jet Lowe, 1984, p. 48
Reservoir Bridge Southwest, Jet Lowe, 1984, p. 49
Riverside Ave. Bridge, Jet Lowe, 1984, p. 64
Rock Island (Government) Bridge,
     J Ceronie, 1985, p. 100
Rouge River Bridge, Jet Lowe, 1990, p. 127
St. Johns Bridge, Jet Lowe, 1990, p. 126
San Francisco–Oakland Bay Bridge,
     Jet Lowe, 1985, p. 142
Scarlets Mill Bridge, Joseph Elliott, 1991
     Christine Ussler, 1991*, p. 79
Schoharie Creek Aqueduct,
     Jack Boucher, 1969, pp. 14–15
Seddon Island Bridge,
     Photographer unknown, 1981, p. 114
Siuslaw River Bridge, Jet Lowe, 1990, pp. 130–131
Smithfield Street Bridge,
     Jack Boucher, 1974, pp. 82–83
Snowden Lift Bridge, Jet Lowe, 1980, pp. 116–117
Starrucca Viaduct, Jack Boucher, 1972, pp. 16–17
Steel Bridge, Jet Lowe, 1990, p. 115
Stewartstown RR Bridge,
     Joseph Elliot, 1991, pp. 60–61
Taft Memorial Bridge, Jet Lowe, 1987, pp. 108–109
Thomas Viaduct, William Barrett, 1970, pp. 10–11
Tunkhannock Viaduct, Jet Lowe, 1982, p. 119
Umpqua River Bridge, Jet Lowe, 1990, p. 132
Upper Pacific Mills Bridge, Martin Stupich, 1984
     Wayne Chang, 1991*, p. 51
Verrazano Narrows Bridge,
     Jet Lowe, 1991, pp. 144–145
Walnut Street Bridge, Joseph Elliott, 1991
     Monika Korsos, 1991*, p. 45
West Main Street Bridge, Jack Boucher, 1971
     Carolyn Givens, 1985*, pp. 56–57
Wheeling Suspension Bridge,
     William Barrett, 1976, pp. 36–37
Whipple Cast and Wrought Iron Bowstring Truss
     Bridge, Jack Boucher, 1969, p. 52
White Bowstring Arch Truss Bridge,
     Louise Taft Cawood, 1986, pp. 72–73
Winona Bridge, Clayton B. Fraser, 1985, pp. 96–97
Yaquina Bay Bridge, Jet Lowe, 1990, p. 133